Student Solutions Manual
for McKeague's
Beginning Algebra
A Text/Workbook

Sixth Edition

Ross Rueger
College of the Sequoias

THOMSON

BROOKS/COLE

Australia • Canada • Mexico • Singapore • Spain • United Kingdom • United States

Printed in the United States of America
1 2 3 4 5 6 7 06 05 04 03 02 01

Printer: Phoenix Color Corp

ISBN 0-534-39852-9

For more information about our products, contact us at:
Thomson Learning Academic Resource Center
1-800-423-0563

For permission to use material from this text, contact us by:
Phone: 1-800-730-2214
Fax: 1-800-731-2215
Web: http://www.thomsonrights.com

Brooks/Cole—Thomson Learning
511 Forest Lodge Road
Pacific Grove, CA 93950
USA

Asia
Thomson Learning
5 Shenton Way #01-01
UIC Building
Singapore 068808

Australia/New Zealand
Thomson Learning
102 Dodds Street
Southbank, Victoria
Australia 3006

Canada
Nelson
1120 Birchmount Road
Toronto, Ontario M1K 5G4
Canada

Europe/Middle East/South Africa
Thomson Learning
High Holborn House
50/51 Bedford Row
London WC1R 4LR
United Kingdom

Latin America
Thomson Learning
Seneca, 53
Colonia Polanco
11560 Mexico D.F.
Mexico

Spain
Paraninfo
Calle/Magallanes, 25
28015 Madrid
Spain

Contents

Preface

This *Student Solutions Manual* contains complete solutions to all odd-numbered exercises (and all chapter test exercises) of *Beginning Algebra: A Text/Workbook* by Charles P. McKeague. I have attempted to format solutions for readability and accuracy, and apologize to you for any errors that you may encounter. If you have any comments, suggestions, error corrections, or alternative solutions please feel free to drop me a note or send an email (address below).

Please use this manual with some degree of caution. Be sure that you have attempted a solution, and re-attempted it, before you look it up in this manual. Mathematics can only be learned by *doing*, and not by observing! As you use this manual, do not just read the solution but work it along with the manual, using my solution to check your work. If you use this manual in that fashion then it should be helpful to you in your studying.

I would like to thank a number of people for their assistance in preparing this manual. Thanks go to Rachael Sturgeon, Julie Foster, and Molly Nance at Brooks/Cole Publishing for their valuable assistance and support. Special thanks go to Matt Bourez of College of the Sequoias for his meticulous error-checking of my solutions, and prompt return of my manuscript under tight deadlines.

I wish to express my appreciation to Pat McKeague for asking me to be involved with this textbook. This book provides a complete introduction to beginning algebra, and you will find the text very easy to read and understand. Good luck!

Ross Rueger
College of the Sequoias
matmanross@aol.com

September, 2002

Chapter 1
The Basics

1.1 Notation and Symbols

1. The equivalent expression is $x + 5 = 14$.

3. The equivalent expression is $5y < 30$.

5. The equivalent expression is $3y \le y + 6$.

7. The equivalent expression is $\dfrac{x}{3} = x + 2$.

9. Expanding the expression: $3^2 = 3 \cdot 3 = 9$

11. Expanding the expression: $7^2 = 7 \cdot 7 = 49$

13. Expanding the expression: $2^3 = 2 \cdot 2 \cdot 2 = 8$

15. Expanding the expression: $4^3 = 4 \cdot 4 \cdot 4 = 64$

17. Expanding the expression: $2^4 = 2 \cdot 2 \cdot 2 \cdot 2 = 16$

19. Expanding the expression: $10^2 = 10 \cdot 10 = 100$

21. Expanding the expression: $11^2 = 11 \cdot 11 = 121$

23. Using the order of operations: $2 \cdot 3 + 5 = 6 + 5 = 11$

25. Using the order of operations: $2(3 + 5) = 2(8) = 16$

27. Using the order of operations: $5 + 2 \cdot 6 = 5 + 12 = 17$

29. Using the order of operations: $(5 + 2) \cdot 6 = 7 \cdot 6 = 42$

31. Using the order of operations: $5 \cdot 4 + 5 \cdot 2 = 20 + 10 = 30$

33. Using the order of operations: $5(4 + 2) = 5(6) = 30$

35. Using the order of operations: $8 + 2(5 + 3) = 8 + 2(8) = 8 + 16 = 24$

37. Using the order of operations: $(8 + 2)(5 + 3) = (10)(8) = 80$

39. Using the order of operations: $20 + 2(8 - 5) + 1 = 20 + 2(3) + 1 = 20 + 6 + 1 = 27$

41. Using the order of operations: $5 + 2(3 \cdot 4 - 1) + 8 = 5 + 2(12 - 1) + 8 = 5 + 2(11) + 8 = 5 + 22 + 8 = 35$

43. Using the order of operations: $8 + 10 \div 2 = 8 + 5 = 13$

45. Using the order of operations: $4 + 8 \div 4 - 2 = 4 + 2 - 2 = 4$

47. Using the order of operations: $3 + 12 \div 3 + 6 \cdot 5 = 3 + 4 + 30 = 37$

49. Using the order of operations: $3 \cdot 8 + 10 \div 2 + 4 \cdot 2 = 24 + 5 + 8 = 37$

51. Using the order of operations: $(5 + 3)(5 - 3) = (8)(2) = 16$

53. Using the order of operations: $5^2 - 3^2 = 5 \cdot 5 - 3 \cdot 3 = 25 - 9 = 16$

55. Using the order of operations: $(4 + 5)^2 = 9^2 = 9 \cdot 9 = 81$

57. Using the order of operations: $4^2 + 5^2 = 4 \cdot 4 + 5 \cdot 5 = 16 + 25 = 41$

59. Using the order of operations: $3 \cdot 10^2 + 4 \cdot 10 + 5 = 300 + 40 + 5 = 345$

61. Using the order of operations: $2 \cdot 10^3 + 3 \cdot 10^2 + 4 \cdot 10 + 5 = 2000 + 300 + 40 + 5 = 2345$

63. Using the order of operations: $10 - 2(4 \cdot 5 - 16) = 10 - 2(20 - 16) = 10 - 2(4) = 10 - 8 = 2$

65. Using the order of operations: $4[7 + 3(2 \cdot 9 - 8)] = 4[7 + 3(18 - 8)] = 4[7 + 3(10)] = 4(7 + 30) = 4(37) = 148$

67. Using the order of operations: $5(7 - 3) + 8(6 - 4) = 5(4) + 8(2) = 20 + 16 = 36$

69. Using the order of operations: $3(4 \cdot 5 - 12) + 6(7 \cdot 6 - 40) = 3(20 - 12) + 6(42 - 40) = 3(8) + 6(2) = 24 + 12 = 36$

71. Using the order of operations: $3^4 + 4^2 \div 2^3 - 5^2 = 81 + 16 \div 8 - 25 = 81 + 2 - 25 = 58$

73. Using the order of operations: $5^2 + 3^4 \div 9^2 + 6^2 = 25 + 81 \div 81 + 36 = 25 + 1 + 36 = 62$

75. There are $5 \cdot 2 = 10$ cookies in the package.

77.   The total number of calories is $210 \cdot 2 = 420$ calories.

79.   There are $7 \cdot 32 = 224$ chips in the bag.

81.   For a person eating 3,000 calories per day, the recommended amount of fat would be $80 + 15 = 95$ grams.

83.   **a.**   The amount of caffeine is: $6(100) = 600$ mg

   b.   The amount of caffeine is: $2(45) + 3(47) = 90 + 141 = 231$ mg

85.   Completing the table:

Activity	Calories burned in 1 hour
Bicycling	374
Bowling	265
Handball	680
Jogging	680
Skiing	544

87.   The next number is 5.

89.   The next number is 10.

91.   The next number is $5^2 = 25$.

93.   Since $2 + 2 = 4$ and $2 + 4 = 6$, the next number is $4 + 6 = 10$.

1.2 Real Numbers

1.   Labeling the point:

3.   Labeling the point:

5.   Labeling the point:

7.   Labeling the point:

9.   Building the fraction: $\frac{3}{4} = \frac{3}{4} \cdot \frac{6}{6} = \frac{18}{24}$

11.   Building the fraction: $\frac{1}{2} = \frac{1}{2} \cdot \frac{12}{12} = \frac{12}{24}$

13.   Building the fraction: $\frac{5}{8} = \frac{5}{8} \cdot \frac{3}{3} = \frac{15}{24}$

15.   Building the fraction: $\frac{3}{5} = \frac{3}{5} \cdot \frac{12}{12} = \frac{36}{60}$

17.   Building the fraction: $\frac{11}{30} = \frac{11}{30} \cdot \frac{2}{2} = \frac{22}{60}$

19.   The opposite of 10 is –10, the reciprocal is $\frac{1}{10}$, and the absolute value is $|10| = 10$.

21.   The opposite of $\frac{3}{4}$ is $-\frac{3}{4}$, the reciprocal is $\frac{4}{3}$, and the absolute value is $\left|\frac{3}{4}\right| = \frac{3}{4}$.

23.   The opposite of $\frac{11}{2}$ is $-\frac{11}{2}$, the reciprocal is $\frac{2}{11}$, and the absolute value is $\left|\frac{11}{2}\right| = \frac{11}{2}$.

25.   The opposite of –3 is 3, the reciprocal is $-\frac{1}{3}$, and the absolute value is $|-3| = 3$.

27.   The opposite of $-\frac{2}{5}$ is $\frac{2}{5}$, the reciprocal is $-\frac{5}{2}$, and the absolute value is $\left|-\frac{2}{5}\right| = \frac{2}{5}$.

29.   The opposite of x is $-x$, the reciprocal is $\frac{1}{x}$, and the absolute value is $|x|$.

31.   The correct symbol is <: $-5 < -3$ **33.**   The correct symbol is >: $-3 > -7$

35.   Since $|-4| = 4$ and $-|-4| = -4$, the correct symbol is >: $|-4| > -|-4|$

37.   Since $-|-7| = -7$, the correct symbol is >: $7 > -|-7|$

39.   The correct symbol is <: $-\frac{3}{4} < -\frac{1}{4}$ **41.**   The correct symbol is <: $-\frac{3}{2} < -\frac{3}{4}$

43.   Simplifying the expression: $|8 - 2| = |6| = 6$

45.   Simplifying the expression: $|5 \cdot 2^3 - 2 \cdot 3^2| = |5 \cdot 8 - 2 \cdot 9| = |40 - 18| = |22| = 22$

47.   Simplifying the expression: $|7 - 2| - |4 - 2| = |5| - |2| = 5 - 2 = 3$

49.   Simplifying the expression: $10 - |7 - 2(5 - 3)| = 10 - |7 - 2(2)| = 10 - |7 - 4| = 10 - |3| = 10 - 3 = 7$

51. Simplifying the expression:
$$15 - |8 - 2(3 \cdot 4 - 9)| - 10 = 15 - |8 - 2(12 - 9)| - 10$$
$$= 15 - |8 - 2(3)| - 10$$
$$= 15 - |8 - 6| - 10$$
$$= 15 - |2| - 10$$
$$= 15 - 2 - 10$$
$$= 3$$

53. Multiplying the fractions: $\frac{2}{3} \cdot \frac{4}{5} = \frac{8}{15}$

55. Multiplying the fractions: $\frac{1}{2}(3) = \frac{1}{2} \cdot \frac{3}{1} = \frac{3}{2}$

57. Multiplying the fractions: $\frac{1}{4}(5) = \frac{1}{4} \cdot \frac{5}{1} = \frac{5}{4}$

59. Multiplying the fractions: $\frac{4}{3} \cdot \frac{3}{4} = \frac{12}{12} = 1$

61. Multiplying the fractions: $6\left(\frac{1}{6}\right) = \frac{6}{1} \cdot \frac{1}{6} = \frac{6}{6} = 1$

63. Multiplying the fractions: $3 \cdot \frac{1}{3} = \frac{3}{1} \cdot \frac{1}{3} = \frac{3}{3} = 1$

65. Expanding the exponent: $\left(\frac{3}{4}\right)^2 = \frac{3}{4} \cdot \frac{3}{4} = \frac{9}{16}$

67. Expanding the exponent: $\left(\frac{2}{3}\right)^3 = \frac{2}{3} \cdot \frac{2}{3} \cdot \frac{2}{3} = \frac{8}{27}$

69. Expanding the exponent: $\left(\frac{1}{10}\right)^4 = \frac{1}{10} \cdot \frac{1}{10} \cdot \frac{1}{10} \cdot \frac{1}{10} = \frac{1}{10,000}$

71. The next number is $\frac{1}{9}$.

73. The next number is $\frac{1}{5^2} = \frac{1}{25}$.

75. The perimeter is $4(1 \text{ in.}) = 4 \text{ in.}$, and the area is $(1 \text{ in.})^2 = 1 \text{ in.}^2$.

77. The perimeter is $2(1.5 \text{ in.}) + 2(0.75 \text{ in.}) = 3.0 \text{ in.} + 1.5 \text{ in.} = 4.5 \text{ in.}$, and the area is $(1.5 \text{ in.})(0.75 \text{ in.}) = 1.125 \text{ in.}^2$.

79. The perimeter is $2.75 \text{ cm} + 4 \text{ cm} + 3.5 \text{ cm} = 10.25 \text{ cm}$, and the area is $\frac{1}{2}(4 \text{ cm})(2.5 \text{ cm}) = 5.0 \text{ cm}^2$.

81. A loss of 8 yards corresponds to –8 on a number line. The total yards gained corresponds to –2 yards.

83. The temperature can be represented as –64°. The new (warmer) temperature corresponds to –54°.

85. The wind chill temperature is –15°.

87. Completing the table:

Month	Temperature (°F)
January	−36
February	−30
March	−14
April	−2
May	19
June	22
July	35
August	30
September	19
October	15
November	−11
December	−26

89. His position corresponds to –100 feet. His new (deeper) position corresponds to –105 feet.

91. The area is given by: $8\frac{1}{2} \cdot 11 = \frac{17}{2} \cdot \frac{11}{1} = \frac{187}{2} = 93.5 \text{ in.}^2$

The perimeter is given by: $2\left(8\frac{1}{2}\right) + 2(11) = 17 + 22 = 39 \text{ in.}$

93. The calories consumed would be: $2(544) + 299 = 1,387$ calories

95. The calories consumed by the 180 lb person would be $3(653) = 1,959$ calories, while the calories consumed by the 120 lb person would be $3(435) = 1,305$ calories. Thus the 180 lb person consumed $1959 - 1306 = 654$ more calories.

1.3 Addition of Real Numbers

1. Adding all positive and negative combinations of 3 and 5:
$$3+5=8$$
$$3+(-5)=-2$$
$$-3+5=2$$
$$(-3)+(-5)=-8$$

3. Adding all positive and negative combinations of 15 and 20:
$$15+20=35$$
$$15+(-20)=-5$$
$$-15+20=5$$
$$(-15)+(-20)=-35$$

5. Adding the numbers: $6+(-3)=3$ **7.** Adding the numbers: $13+(-20)=-7$

9. Adding the numbers: $18+(-32)=-14$ **11.** Adding the numbers: $-6+3=-3$

13. Adding the numbers: $-30+5=-25$ **15.** Adding the numbers: $-6+(-6)=-12$

17. Adding the numbers: $-9+(-10)=-19$ **19.** Adding the numbers: $-10+(-15)=-25$

21. Performing the additions: $5+(-6)+(-7)=5+(-13)=-8$

23. Performing the additions: $-7+8+(-5)=-12+8=-4$

25. Performing the additions: $5+[6+(-2)]+(-3)=5+4+(-3)=9+(-3)=6$

27. Performing the additions: $[6+(-2)]+[3+(-1)]=4+2=6$

29. Performing the additions: $20+(-6)+[3+(-9)]=20+(-6)+(-6)=20+(-12)=8$

31. Performing the additions: $-3+(-2)+[5+(-4)]=-3+(-2)+1=-5+1=-4$

33. Performing the additions: $(-9+2)+[5+(-8)]+(-4)=-7+(-3)+(-4)=-14$

35. Performing the additions: $[-6+(-4)]+[7+(-5)]+(-9)=-10+2+(-9)=-19+2=-17$

37. Performing the additions: $(-6+9)+(-5)+(-4+3)+7=3+(-5)+(-1)+7=10+(-6)=4$

39. Using order of operations: $-5+2(-3+7)=-5+2(4)=-5+8=3$

41. Using order of operations: $9+3(-8+10)=9+3(2)=9+6=15$

43. Using order of operations: $-10+2(-6+8)+(-2)=-10+2(2)+(-2)=-10+4+(-2)=-12+4=-8$

45. Using order of operations: $2(-4+7)+3(-6+8)=2(3)+3(2)=6+6=12$

47. The pattern is to add 5, so the next two terms are $18+5=23$ and $23+5=28$.

49. The pattern is to add 5, so the next two terms are $25+5=30$ and $30+5=35$.

51. The pattern is to add -5, so the next two terms are $5+(-5)=0$ and $0+(-5)=-5$.

53. The pattern is to add -6, so the next two terms are $-6+(-6)=-12$ and $-12+(-6)=-18$.

55. The pattern is to add -4, so the next two terms are $0+(-4)=-4$ and $-4+(-4)=-8$.

57. Yes, since each successive odd number is 2 added to the previous one.

59. The expression is: $5+9=14$ **61.** The expression is: $[-7+(-5)]+4=-12+4=-8$

63. The expression is: $[-2+(-3)]+10=-5+10=5$ **65.** The number is 3, since $-8+3=-5$.

67. The number is -3, since $-6+(-3)=-9$.   **69.** The expression is $-12°+4°=-8°$.

71. The expression is $\$10+(-\$6)+(-\$8)=\$10+(-\$14)=-\$4$.

73. The new balance is $-\$30+\$40=\$10$.

75. **a.** Completing the table:

Profession	Salary
Engineering	$42,862
Computer Science	$40,920
Math / Statistics	$40,523
Chemistry	$36,036
Business Administration	$34,831
Accounting	$33,702
Sales / Marketing	$33,252
Teaching	$25,735

   b. The average family income is: $\$36,036+\$42,862=\$78,898$

1.4 Subtraction of Real Numbers

1. Subtracting the numbers: $5 - 8 = 5 + (-8) = -3$

3. Subtracting the numbers: $3 - 9 = 3 + (-9) = -6$

5. Subtracting the numbers: $5 - 5 = 5 + (-5) = 0$

7. Subtracting the numbers: $-8 - 2 = -8 + (-2) = -10$

9. Subtracting the numbers: $-4 - 12 = -4 + (-12) = -16$

11. Subtracting the numbers: $-6 - 6 = -6 + (-6) = -12$

13. Subtracting the numbers: $-8 - (-1) = -8 + 1 = -7$

15. Subtracting the numbers: $15 - (-20) = 15 + 20 = 35$

17. Subtracting the numbers: $-4 - (-4) = -4 + 4 = 0$

19. Using order of operations: $3 - 2 - 5 = 3 + (-2) + (-5) = 3 + (-7) = -4$

21. Using order of operations: $9 - 2 - 3 = 9 + (-2) + (-3) = 9 + (-5) = 4$

23. Using order of operations: $-6 - 8 - 10 = -6 + (-8) + (-10) = -24$

25. Using order of operations: $-22 + 4 - 10 = -22 + 4 + (-10) = -32 + 4 = -28$

27. Using order of operations: $10 - (-20) - 5 = 10 + 20 + (-5) = 30 + (-5) = 25$

29. Using order of operations: $8 - (2 - 3) - 5 = 8 - (-1) - 5 = 8 + 1 + (-5) = 9 + (-5) = 4$

31. Using order of operations: $7 - (3 - 9) - 6 = 7 - (-6) - 6 = 7 + 6 + (-6) = 13 + (-6) = 7$

33. Using order of operations: $5 - (-8 - 6) - 2 = 5 - (-14) - 2 = 5 + 14 + (-2) = 19 + (-2) = 17$

35. Using order of operations: $-(5 - 7) - (2 - 8) = -(-2) - (-6) = 2 + 6 = 8$

37. Using order of operations: $-(3 - 10) - (6 - 3) = -(-7) - 3 = 7 + (-3) = 4$

39. Using order of operations: $16 - [(4 - 5) - 1] = 16 - (-1 - 1) = 16 - (-2) = 16 + 2 = 18$

41. Using order of operations: $5 - [(2 - 3) - 4] = 5 - (-1 - 4) = 5 - (-5) = 5 + 5 = 10$

43. Using order of operations:
$$21 - [-(3 - 4) - 2] - 5 = 21 - [-(-1) - 2] - 5 = 21 - (1 - 2) - 5 = 21 - (-1) - 5 = 21 + 1 + (-5) = 22 + (-5) = 17$$

45. Using order of operations: $2 \cdot 8 - 3 \cdot 5 = 16 - 15 = 16 + (-15) = 1$

47. Using order of operations: $3 \cdot 5 - 2 \cdot 7 = 15 - 14 = 15 + (-14) = 1$

49. Using order of operations: $5 \cdot 9 - 2 \cdot 3 - 6 \cdot 2 = 45 - 6 - 12 = 45 + (-6) + (-12) = 45 + (-18) = 27$

51. Using order of operations: $3 \cdot 8 - 2 \cdot 4 - 6 \cdot 7 = 24 - 8 - 42 = 24 + (-8) + (-42) = 24 + (-50) = -26$

53. Using order of operations: $2 \cdot 3^2 - 5 \cdot 2^2 = 2 \cdot 9 - 5 \cdot 4 = 18 - 20 = 18 + (-20) = -2$

55. Using order of operations: $4 \cdot 3^3 - 5 \cdot 2^3 = 4 \cdot 27 - 5 \cdot 8 = 108 - 40 = 108 + (-40) = 68$

57. Writing the expression: $-7 - 4 = -7 + (-4) = -11$

59. Writing the expression: $12 - (-8) = 12 + 8 = 20$

61. Writing the expression: $-5 - (-7) = -5 + 7 = 2$

63. Writing the expression: $[4 + (-5)] - 17 = -1 - 17 = -1 + (-17) = -18$

65. Writing the expression: $8 - 5 = 8 + (-5) = 3$

67. Writing the expression: $-8 - 5 = -8 + (-5) = -13$

69. Writing the expression: $8 - (-5) = 8 + 5 = 13$

71. The number is 10, since $8 - 10 = 8 + (-10) = -2$.

73. The number is -2, since $8 - (-2) = 8 + 2 = 10$.

75. The expression is $\$1,500 - \$730 = \$770$.

77. The expression is $-\$35 + \$15 - \$20 = -\$35 + (-\$20) + \$15 = -\$55 + \$15 = -\$40$.

79. The expression is $\$98 - \$65 - \$53 = \$98 + (-\$65) + (-\$53) = \$98 + (-\$118) = -\$20$.

81. The sequence of values is $\$4500$, $\$3950$, $\$3400$, $\$2850$, and $\$2300$. This is an arithmetic sequence, since $-\$550$ is added to each value to obtain the new value.

83. The difference is $1000 \text{ feet} - 231 \text{ feet} = 769 \text{ feet}$.

85. He is 439 feet from the starting line.

87. 2 seconds have gone by.

89. The angles add to $90°$, so $x = 90° - 55° = 35°$.

91. The angles add to $180°$, so $x = 180° - 120° = 60°$.

93. **a.** Completing the table:

Year	Garbage (Millions of tons)
1960	88
1970	121
1980	152
1990	205
1997	217

b. Subtracting: $205 - 121 = 84$. There was 84 million tons more garbage in 1990 than in 1970.

95. **a.**   Completing the table:

Year	Cost (cents / minute)
1998	33
1999	28
2000	25
2001	23
2002	22
2003	20

      b.   The difference is cost is $33 - 28 = 5$ cents/minute.

1.5 Properties of Real Numbers

1.   commutative property (of addition)

3.   multiplicative inverse property

5.   commutative property (of addition)

7.   distributive property

9.   commutative and associative properties (of addition)

11.   commutative and associative properties (of addition)

13.   commutative property (of addition)

15.   commutative and associative properties (of multiplication)

17.   commutative property (of multiplication)

19.   additive inverse property

21.   The expression should read $3(x+2) = 3x + 6$.

23.   The expression should read $9(a+b) = 9a + 9b$.

25.   The expression should read $3(0) = 0$.

27.   The expression should read $3 + (-3) = 0$.

29.   The expression should read $10(1) = 10$.

31.   Simplifying the expression: $4 + (2 + x) = (4 + 2) + x = 6 + x$

33.   Simplifying the expression: $(x + 2) + 7 = x + (2 + 7) = x + 9$

35.   Simplifying the expression: $3(5x) = (3 \cdot 5)x = 15x$

37.   Simplifying the expression: $9(6y) = (9 \cdot 6)y = 54y$

39.   Simplifying the expression: $\frac{1}{2}(3a) = \left(\frac{1}{2} \cdot 3\right)a = \frac{3}{2}a$

41.   Simplifying the expression: $\frac{1}{3}(3x) = \left(\frac{1}{3} \cdot 3\right)x = 1x = x$

43.   Simplifying the expression: $\frac{1}{2}(2y) = \left(\frac{1}{2} \cdot 2\right)y = 1y = y$

45.   Simplifying the expression: $\frac{3}{4}\left(\frac{4}{3}x\right) = \left(\frac{3}{4} \cdot \frac{4}{3}\right)x = 1x = x$

47.   Simplifying the expression: $\frac{6}{5}\left(\frac{5}{6}a\right) = \left(\frac{6}{5} \cdot \frac{5}{6}\right)a = 1a = a$

49.   Applying the distributive property: $8(x + 2) = 8 \cdot x + 8 \cdot 2 = 8x + 16$

51.   Applying the distributive property: $8(x - 2) = 8 \cdot x - 8 \cdot 2 = 8x - 16$

53.   Applying the distributive property: $4(y + 1) = 4 \cdot y + 4 \cdot 1 = 4y + 4$

55.   Applying the distributive property: $3(6x + 5) = 3 \cdot 6x + 3 \cdot 5 = 18x + 15$

57.   Applying the distributive property: $2(3a + 7) = 2 \cdot 3a + 2 \cdot 7 = 6a + 14$

59.   Applying the distributive property: $9(6y - 8) = 9 \cdot 6y - 9 \cdot 8 = 54y - 72$

61.   Applying the distributive property: $\frac{1}{2}(3x - 6) = \frac{1}{2} \cdot 3x - \frac{1}{2} \cdot 6 = \frac{3}{2}x - 3$

63.   Applying the distributive property: $\frac{1}{3}(3x + 6) = \frac{1}{3} \cdot 3x + \frac{1}{3} \cdot 6 = x + 2$

65.   Applying the distributive property: $3(x + y) = 3x + 3y$

67.   Applying the distributive property: $8(a - b) = 8a - 8b$

69.   Applying the distributive property: $6(2x + 3y) = 6 \cdot 2x + 6 \cdot 3y = 12x + 18y$

71.   Applying the distributive property: $4(3a - 2b) = 4 \cdot 3a - 4 \cdot 2b = 12a - 8b$

73.   Applying the distributive property: $\frac{1}{2}(6x + 4y) = \frac{1}{2} \cdot 6x + \frac{1}{2} \cdot 4y = 3x + 2y$

75.   Applying the distributive property: $4(a + 4) + 9 = 4a + 16 + 9 = 4a + 25$

77.   Applying the distributive property: $2(3x + 5) + 2 = 6x + 10 + 2 = 6x + 12$

79.   Applying the distributive property: $7(2x + 4) + 10 = 14x + 28 + 10 = 14x + 38$

81.   No. The man cannot reverse the order of putting on his socks and putting on his shoes.

83.   No. The skydiver must jump out of the plane before pulling the rip cord.

85.   Division is not a commutative operation. For example, $8 \div 4 = 2$ while $4 \div 8 = \frac{1}{2}$.

87.   Computing the yearly take-home pay:
$$12(2400 - 480) = 12(1920) = \$23{,}040 \qquad 12 \cdot 2400 - 12 \cdot 480 = 28{,}800 - 5{,}760 = \$23{,}040$$

89.   Rewriting the formula: $P = 2w + 2l = 2(w + l)$

1.6 Multiplication of Real Numbers

1. Finding the product: $7(-6) = -42$

3. Finding the product: $-8(2) = -16$

5. Finding the product: $-3(-1) = 3$

7. Finding the product: $-11(-11) = 121$

9. Using order of operations: $-3(2)(-1) = 6$

11. Using order of operations: $-3(-4)(-5) = -60$

13. Using order of operations: $-2(-4)(-3)(-1) = 24$

15. Using order of operations: $(-7)^2 = (-7)(-7) = 49$

17. Using order of operations: $(-3)^3 = (-3)(-3)(-3) = -27$

19. Using order of operations: $-2(2-5) = -2(-3) = 6$

21. Using order of operations: $-5(8-10) = -5(-2) = 10$

23. Using order of operations: $(4-7)(6-9) = (-3)(-3) = 9$

25. Using order of operations: $(-3-2)(-5-4) = (-5)(-9) = 45$

27. Using order of operations: $-3(-6) + 4(-1) = 18 + (-4) = 14$

29. Using order of operations: $2(3) - 3(-4) + 4(-5) = 6 + 12 + (-20) = 18 + (-20) = -2$

31. Using order of operations: $4(-3)^2 + 5(-6)^2 = 4(9) + 5(36) = 36 + 180 = 216$

33. Using order of operations: $7(-2)^3 - 2(-3)^3 = 7(-8) - 2(-27) = -56 + 54 = -2$

35. Using order of operations: $6 - 4(8-2) = 6 - 4(6) = 6 - 24 = 6 + (-24) = -18$

37. Using order of operations: $9 - 4(3-8) = 9 - 4(-5) = 9 + 20 = 29$

39. Using order of operations: $-4(3-8) - 6(2-5) = -4(-5) - 6(-3) = 20 + 18 = 38$

41. Using order of operations: $7 - 2[-6 - 4(-3)] = 7 - 2(-6 + 12) = 7 - 2(6) = 7 - 12 = 7 + (-12) = -5$

43. Using order of operations:
$$7 - 3[2(-4-4) - 3(-1-1)] = 7 - 3[2(-8) - 3(-2)] = 7 - 3(-16 + 6) = 7 - 3(-10) = 7 + 30 = 37$$

45. Using order of operations:
$$8 - 6[-2(-3-1) + 4(-2-3)] = 8 - 6[-2(-4) + 4(-5)] = 8 - 6[8 + (-20)] = 8 - 6(-12) = 8 + 72 = 80$$

47. Multiplying the fractions: $-\frac{2}{3} \cdot \frac{5}{7} = -\frac{2 \cdot 5}{3 \cdot 7} = -\frac{10}{21}$

49. Multiplying the fractions: $-8\left(\frac{1}{2}\right) = -\frac{8}{1} \cdot \frac{1}{2} = -\frac{8}{2} = -4$

51. Multiplying the fractions: $-\frac{3}{4}\left(-\frac{4}{3}\right) = -\frac{3}{4} \cdot \left(-\frac{4}{3}\right) = \frac{12}{12} = 1$

53. Multiplying the fractions: $\left(-\frac{3}{4}\right)^2 = \left(-\frac{3}{4}\right)\left(-\frac{3}{4}\right) = \frac{9}{16}$

55. Multiplying the expressions: $-2(4x) = (-2 \cdot 4)x = -8x$

57. Multiplying the expressions: $-7(-6x) = [-7 \cdot (-6)]x = 42x$

59. Multiplying the expressions: $-\frac{1}{3}(-3x) = \left[-\frac{1}{3} \cdot (-3)\right]x = 1x = x$

61. Simplifying the expression: $-4(a+2) = -4a + (-4)(2) = -4a - 8$

63. Simplifying the expression: $-\frac{1}{2}(3x-6) = -\frac{3}{2}x - \frac{1}{2}(-6) = -\frac{3}{2}x + 3$

65. Simplifying the expression: $-3(2x-5) - 7 = -6x + 15 - 7 = -6x + 8$

67. Simplifying the expression: $-5(3x+4) - 10 = -15x - 20 - 10 = -15x - 30$

69. Writing the expression: $3(-10) + 5 = -30 + 5 = -25$

71. Writing the expression: $2(-4x) = -8x$

73. Writing the expression: $-9 \cdot 2 - 8 = -18 + (-8) = -26$

75. The pattern is to multiply by 2, so the next number is $4 \cdot 2 = 8$.

77. The pattern is to multiply by –2, so the next number is $40 \cdot (-2) = -80$.

79. The pattern is to multiply by $\frac{1}{2}$, so the next number is $\frac{1}{4} \cdot \frac{1}{2} = \frac{1}{8}$.

81. The pattern is to multiply by –2, so the next number is $12 \cdot (-2) = -24$.

83. The amount lost is: $20(\$3) = \$60$

85. The temperature is: $25° - 4(6°) = 25° - 24° = 1°$

87. The sequence of values is $500, $1000, $2000, $4000, $8000, and $16000. Yes, this is a geometric sequence, since each value is 2 times the preceding value.

89. The net change in calories is: $2(630) - 3(265) = 1260 - 795 = 465$ calories

1.7 Division of Real Numbers

1. Finding the quotient: $\dfrac{8}{-4} = -2$

3. Finding the quotient: $\dfrac{-48}{16} = -3$

5. Finding the quotient: $\dfrac{-7}{21} = -\frac{1}{3}$

7. Finding the quotient: $\dfrac{-39}{-13} = 3$

9. Finding the quotient: $\dfrac{-6}{-42} = \frac{1}{7}$

11. Finding the quotient: $\dfrac{0}{-32} = 0$

13. Performing the operations: $-3 + 12 = 9$

15. Performing the operations: $-3 - 12 = -3 + (-12) = -15$

17. Performing the operations: $-3(12) = -36$

19. Performing the operations: $-3 \div 12 = \dfrac{-3}{12} = -\frac{1}{4}$

21. Dividing and reducing: $\frac{4}{5} \div \frac{3}{4} = \frac{4}{5} \cdot \frac{4}{3} = \frac{16}{15}$

23. Dividing and reducing: $-\frac{5}{6} \div \left(-\frac{5}{8}\right) = -\frac{5}{6} \cdot \left(-\frac{8}{5}\right) = \frac{40}{30} = \frac{4}{3}$

25. Dividing and reducing: $\frac{10}{13} \div \left(-\frac{5}{4}\right) = \frac{10}{13} \cdot \left(-\frac{4}{5}\right) = -\frac{40}{65} = -\frac{8}{13}$

27. Dividing and reducing: $-\frac{5}{6} \div \frac{5}{6} = -\frac{5}{6} \cdot \frac{6}{5} = -\frac{30}{30} = -1$

29. Dividing and reducing: $-\frac{3}{4} \div \left(-\frac{3}{4}\right) = -\frac{3}{4} \cdot \left(-\frac{4}{3}\right) = \frac{12}{12} = 1$

31. Using order of operations: $\dfrac{3(-2)}{-10} = \dfrac{-6}{-10} = \frac{3}{5}$

33. Using order of operations: $\dfrac{-5(-5)}{-15} = \dfrac{25}{-15} = -\frac{5}{3}$

35. Using order of operations: $\dfrac{-8(-7)}{-28} = \dfrac{56}{-28} = -2$

37. Using order of operations: $\dfrac{27}{4-13} = \dfrac{27}{-9} = -3$

39. Using order of operations: $\dfrac{20-6}{5-5} = \dfrac{14}{0} = \text{undefined}$

41. Using order of operations: $\dfrac{-3+9}{2 \cdot 5 - 10} = \dfrac{6}{10-10} = \dfrac{6}{0} = \text{undefined}$

43. Using order of operations: $\dfrac{15(-5)-25}{2(-10)} = \dfrac{-75-25}{-20} = \dfrac{-100}{-20} = 5$

45. Using order of operations: $\dfrac{27-2(-4)}{-3(5)} = \dfrac{27+8}{-15} = \dfrac{35}{-15} = -\frac{7}{3}$

47. Using order of operations: $\dfrac{12-6(-2)}{12(-2)} = \dfrac{12+12}{-24} = \dfrac{24}{-24} = -1$

49. Using order of operations: $\dfrac{5^2-2^2}{-5+2} = \dfrac{25-4}{-3} = \dfrac{21}{-3} = -7$

51. Using order of operations: $\dfrac{8^2-2^2}{8^2+2^2} = \dfrac{64-4}{64+4} = \dfrac{60}{68} = \frac{15}{17}$

53. Using order of operations: $\dfrac{(5+3)^2}{-5^2-3^2} = \dfrac{8^2}{-25-9} = \dfrac{64}{-34} = -\frac{32}{17}$

55. Using order of operations: $\dfrac{(8-4)^2}{8^2-4^2} = \dfrac{4^2}{64-16} = \dfrac{16}{48} = \frac{1}{3}$

57. Using order of operations: $\dfrac{-4 \cdot 3^2 - 5 \cdot 2^2}{-8(7)} = \dfrac{-4 \cdot 9 - 5 \cdot 4}{-56} = \dfrac{-36-20}{-56} = \dfrac{-56}{-56} = 1$

59. Using order of operations: $\dfrac{3 \cdot 10^2 + 4 \cdot 10 + 5}{345} = \dfrac{300+40+5}{345} = \dfrac{345}{345} = 1$

61. Using order of operations: $\dfrac{7-[(2-3)-4]}{-1-2-3} = \dfrac{7-(-1-4)}{-6} = \dfrac{7-(-5)}{-6} = \dfrac{7+5}{-6} = \dfrac{12}{-6} = -2$

63. Using order of operations: $\dfrac{6(-4)-2(5-8)}{-6-3-5} = \dfrac{-24-2(-3)}{-14} = \dfrac{-24+6}{-14} = \dfrac{-18}{-14} = \frac{9}{7}$

65. Using order of operations: $\dfrac{3(-5-3)+4(7-9)}{5(-2)+3(-4)} = \dfrac{3(-8)+4(-2)}{-10+(-12)} = \dfrac{-24+(-8)}{-22} = \dfrac{-32}{-22} = \frac{16}{11}$

67. Using order of operations: $\dfrac{|3-9|}{3-9} = \dfrac{|-6|}{-6} = \dfrac{6}{-6} = -1$ 69. The quotient is $\dfrac{-12}{-4} = 3$.

71. The number is –10, since $\dfrac{-10}{-5} = 2$. 73. The number is –3, since $\dfrac{27}{-3} = -9$.

75. The expression is: $\dfrac{-20}{4} - 3 = -5 - 3 = -8$

77. Each person would lose: $\dfrac{13600 - 15000}{4} = \dfrac{-1400}{4} = -350 = \$350$ loss

79. The change per hour is: $\dfrac{61° - 75°}{4} = \dfrac{-14°}{4} = -3.5°$ per hour

81. For the year 2000, the profit was: $\$11,500 - \$9,500 = \$2,000$

83. In the year 2000, the largest increase in revenue was: $\$11,500 - \$7,750 = \$3,750$

1.8 Subsets of the Real Numbers

1. The whole numbers are: 0, 1

3. The rational numbers are: $-3, -2.5, 0, 1, \frac{3}{2}$

5. The real numbers are: $-3, -2.5, 0, 1, \frac{3}{2}, \sqrt{15}$

7. The integers are: –10, –8, –2, 9

9. The irrational numbers are: π

11. true

13. false

15. false

17. true

19. This number is composite: $48 = 6 \cdot 8 = (2 \cdot 3) \cdot (2 \cdot 2 \cdot 2) = 2^4 \cdot 3$

21. This number is prime.

23. This number is composite: $1023 = 3 \cdot 341 = 3 \cdot 11 \cdot 31$

25. Factoring the number: $144 = 12 \cdot 12 = (3 \cdot 4) \cdot (3 \cdot 4) = (3 \cdot 2 \cdot 2) \cdot (3 \cdot 2 \cdot 2) = 2^4 \cdot 3^2$

27. Factoring the number: $38 = 2 \cdot 19$

29. Factoring the number: $105 = 5 \cdot 21 = 5 \cdot (3 \cdot 7) = 3 \cdot 5 \cdot 7$

31. Factoring the number: $180 = 10 \cdot 18 = (2 \cdot 5) \cdot (3 \cdot 6) = (2 \cdot 5) \cdot (3 \cdot 2 \cdot 3) = 2^2 \cdot 3^2 \cdot 5$

33. Factoring the number: $385 = 5 \cdot 77 = 5 \cdot (7 \cdot 11) = 5 \cdot 7 \cdot 11$

35. Factoring the number: $121 = 11 \cdot 11 = 11^2$

37. Factoring the number: $420 = 10 \cdot 42 = (2 \cdot 5) \cdot (7 \cdot 6) = (2 \cdot 5) \cdot (7 \cdot 2 \cdot 3) = 2^2 \cdot 3 \cdot 5 \cdot 7$

39. Factoring the number: $620 = 10 \cdot 62 = (2 \cdot 5) \cdot (2 \cdot 31) = 2^2 \cdot 5 \cdot 31$

41. Reducing the fraction: $\dfrac{105}{165} = \dfrac{3 \cdot 5 \cdot 7}{3 \cdot 5 \cdot 11} = \dfrac{7}{11}$

43. Reducing the fraction: $\dfrac{525}{735} = \dfrac{3 \cdot 5 \cdot 5 \cdot 7}{3 \cdot 5 \cdot 7 \cdot 7} = \dfrac{5}{7}$

45. Reducing the fraction: $\dfrac{385}{455} = \dfrac{5 \cdot 7 \cdot 11}{5 \cdot 7 \cdot 13} = \dfrac{11}{13}$

47. Reducing the fraction: $\dfrac{322}{345} = \dfrac{2 \cdot 7 \cdot 23}{3 \cdot 5 \cdot 23} = \dfrac{2 \cdot 7}{3 \cdot 5} = \dfrac{14}{15}$

49. Reducing the fraction: $\dfrac{205}{369} = \dfrac{5 \cdot 41}{3 \cdot 3 \cdot 41} = \dfrac{5}{3 \cdot 3} = \dfrac{5}{9}$

51. Reducing the fraction: $\dfrac{215}{344} = \dfrac{5 \cdot 43}{2 \cdot 2 \cdot 2 \cdot 43} = \dfrac{5}{2 \cdot 2 \cdot 2} = \dfrac{5}{8}$

53. Factoring into prime numbers: $6^3 = (2 \cdot 3)^3 = 2^3 \cdot 3^3$

55. Factoring into prime numbers: $9^4 \cdot 16^2 = (3 \cdot 3)^4 \cdot (2 \cdot 2 \cdot 2 \cdot 2)^2 = 3^4 \cdot 3^4 \cdot 2^2 \cdot 2^2 \cdot 2^2 \cdot 2^2 = 2^8 \cdot 3^8$

57. Simplifying and factoring: $3 \cdot 8 + 3 \cdot 7 + 3 \cdot 5 = 24 + 21 + 15 = 60 = 6 \cdot 10 = (2 \cdot 3) \cdot (2 \cdot 5) = 2^2 \cdot 3 \cdot 5$

59. They are not a subset of the irrational numbers.

61. 8, 21, and 34 are Fibonacci numbers that are composite numbers.

1.9 Addition and Subtraction with Fractions

1. Combining the fractions: $\frac{3}{6}+\frac{1}{6}=\frac{4}{6}=\frac{2}{3}$

3. Combining the fractions: $\frac{3}{8}-\frac{5}{8}=-\frac{2}{8}=-\frac{1}{4}$

5. Combining the fractions: $-\frac{1}{4}+\frac{3}{4}=\frac{2}{4}=\frac{1}{2}$

7. Combining the fractions: $\frac{x}{3}-\frac{1}{3}=\frac{x-1}{3}$

9. Combining the fractions: $\frac{1}{4}+\frac{2}{4}+\frac{3}{4}=\frac{6}{4}=\frac{3}{2}$

11. Combining the fractions: $\frac{x+7}{2}-\frac{1}{2}=\frac{x+7-1}{2}=\frac{x+6}{2}$

13. Combining the fractions: $\frac{1}{10}-\frac{3}{10}-\frac{4}{10}=-\frac{6}{10}=-\frac{3}{5}$

15. Combining the fractions: $\frac{1}{a}+\frac{4}{a}+\frac{5}{a}=\frac{10}{a}$

17. Combining the fractions: $\frac{1}{8}+\frac{3}{4}=\frac{1}{8}+\frac{3\cdot2}{4\cdot2}=\frac{1}{8}+\frac{6}{8}=\frac{7}{8}$

19. Combining the fractions: $\frac{3}{10}-\frac{1}{5}=\frac{3}{10}-\frac{1\cdot2}{5\cdot2}=\frac{3}{10}-\frac{2}{10}=\frac{1}{10}$

21. Combining the fractions: $\frac{4}{9}+\frac{1}{3}=\frac{4}{9}+\frac{1\cdot3}{3\cdot3}=\frac{4}{9}+\frac{3}{9}=\frac{7}{9}$

23. Combining the fractions: $2+\frac{1}{3}=\frac{2\cdot3}{1\cdot3}+\frac{1}{3}=\frac{6}{3}+\frac{1}{3}=\frac{7}{3}$

25. Combining the fractions: $-\frac{3}{4}+1=-\frac{3}{4}+\frac{1\cdot4}{1\cdot4}=-\frac{3}{4}+\frac{4}{4}=\frac{1}{4}$

27. Combining the fractions: $\frac{1}{2}+\frac{2}{3}=\frac{1\cdot3}{2\cdot3}+\frac{2\cdot2}{3\cdot2}=\frac{3}{6}+\frac{4}{6}=\frac{7}{6}$

29. Combining the fractions: $\frac{5}{12}-\left(-\frac{3}{8}\right)=\frac{5}{12}+\frac{3}{8}=\frac{5\cdot2}{12\cdot2}+\frac{3\cdot3}{8\cdot3}=\frac{10}{24}+\frac{9}{24}=\frac{19}{24}$

31. Combining the fractions: $-\frac{1}{20}+\frac{8}{30}=-\frac{1\cdot3}{20\cdot3}+\frac{8\cdot2}{30\cdot2}=-\frac{3}{60}+\frac{16}{60}=\frac{13}{60}$

33. First factor the denominators to find the LCM:
$$30=2\cdot3\cdot5$$
$$42=2\cdot3\cdot7$$
$$LCM=2\cdot3\cdot5\cdot7=210$$
Combining the fractions: $\frac{17}{30}+\frac{11}{42}=\frac{17\cdot7}{30\cdot7}+\frac{11\cdot5}{42\cdot5}=\frac{119}{210}+\frac{55}{210}=\frac{174}{210}=\frac{2\cdot3\cdot29}{2\cdot3\cdot5\cdot7}=\frac{29}{5\cdot7}=\frac{29}{35}$

35. First factor the denominators to find the LCM:
$$84=2\cdot2\cdot3\cdot7$$
$$90=2\cdot3\cdot3\cdot5$$
$$LCM=2\cdot2\cdot3\cdot3\cdot5\cdot7=1260$$
Combining the fractions: $\frac{25}{84}+\frac{41}{90}=\frac{25\cdot15}{84\cdot15}+\frac{41\cdot14}{90\cdot14}=\frac{375}{1260}+\frac{574}{1260}=\frac{949}{1260}$

37. First factor the denominators to find the LCM:
$$126=2\cdot3\cdot3\cdot7$$
$$180=2\cdot2\cdot3\cdot3\cdot5$$
$$LCM=2\cdot2\cdot3\cdot3\cdot5\cdot7=1260$$
Combining the fractions: $\frac{13}{126}-\frac{13}{180}=\frac{13\cdot10}{126\cdot10}-\frac{13\cdot7}{180\cdot7}=\frac{130}{1260}-\frac{91}{1260}=\frac{39}{1260}=\frac{3\cdot13}{2\cdot2\cdot3\cdot3\cdot5\cdot7}=\frac{13}{2\cdot2\cdot3\cdot3\cdot5\cdot7}=\frac{13}{420}$

39. Combining the fractions: $\frac{3}{4}+\frac{1}{8}+\frac{5}{6}=\frac{3\cdot6}{4\cdot6}+\frac{1\cdot3}{8\cdot3}+\frac{5\cdot4}{6\cdot4}=\frac{18}{24}+\frac{3}{24}+\frac{20}{24}=\frac{41}{24}$

41. Combining the fractions: $\frac{1}{2}+\frac{1}{3}+\frac{1}{4}+\frac{1}{6}=\frac{1\cdot6}{2\cdot6}+\frac{1\cdot4}{3\cdot4}+\frac{1\cdot3}{4\cdot3}+\frac{1\cdot2}{6\cdot2}=\frac{6}{12}+\frac{4}{12}+\frac{3}{12}+\frac{2}{12}=\frac{15}{12}=\frac{5}{4}$

43. The sum is given by: $\frac{3}{7}+2+\frac{1}{9}=\frac{3\cdot9}{7\cdot9}+\frac{2\cdot63}{1\cdot63}+\frac{1\cdot7}{9\cdot7}=\frac{27}{63}+\frac{126}{63}+\frac{7}{63}=\frac{160}{63}$

45. The difference is given by: $\frac{7}{8}-\frac{1}{4}=\frac{7}{8}-\frac{1\cdot2}{4\cdot2}=\frac{7}{8}-\frac{2}{8}=\frac{5}{8}$

47. The pattern is to add $-\frac{1}{3}$, so the fourth term is: $-\frac{1}{3}+\left(-\frac{1}{3}\right)=-\frac{2}{3}$

49.   The pattern is to add $\frac{2}{3}$, so the fourth term is: $\frac{5}{3} + \frac{2}{3} = \frac{7}{3}$

51.   The pattern is to multiply by $\frac{1}{5}$, so the fourth term is: $\frac{1}{25} \cdot \frac{1}{5} = \frac{1}{125}$

53.   The perimeter is: $\frac{3}{8} + \frac{3}{8} + \frac{3}{8} + \frac{3}{8} = \frac{12}{8} = \frac{3}{2} = 1\frac{1}{2}$ feet

55.   The perimeter is: $\frac{4}{5} + \frac{3}{10} + \frac{4}{5} + \frac{3}{10} = \frac{8}{10} + \frac{3}{10} + \frac{8}{10} + \frac{3}{10} = \frac{22}{10} = \frac{11}{5} = 2\frac{1}{5}$ centimeters (cm)

Chapter 1 Review

1.   The expression is: $-7 + (-10) = -17$

3.   The expression is: $(-3 + 12) + 5 = 9 + 5 = 14$

5.   The expression is: $9 - (-3) = 9 + 3 = 12$

7.   The expression is: $(-3)(-7) - 6 = 21 - 6 = 15$

9.   The expression is: $2[-8(3x)] = 2(-24x) = -48x$

11.   The expression is: $\frac{-40}{8} - 7 = -5 - 7 = -5 + (-7) = -12$

13.   Labeling the point:

15.   Labeling the point (note that $\frac{24}{8} = 3$):

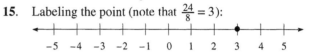

17.   Labeling the point:

1.25

```
 ←——+——+——+——+——+——+——↓——+——+——+——+——→
   -5   -4   -3   -2   -1   0   1   2   3   4   5
```

19.   The absolute value is: $|12| = 12$

21.   The absolute value is: $\left|-\frac{4}{5}\right| = \frac{4}{5}$

23.   Simplifying: $|-1.8| = 1.8$

25.   The opposite is –6, and the reciprocal is $\frac{1}{6}$.

27.   The opposite is 9, and the reciprocal is $-\frac{1}{9}$.

29.   Multiplying the fractions: $\left(\frac{2}{5}\right)\left(\frac{3}{7}\right) = \frac{6}{35}$

31.   Multiplying the fractions: $\left(-\frac{4}{5}\right)\left(\frac{25}{16}\right) = -\frac{100}{80} = -\frac{5}{4}$

33.   Adding: $-18 + (-20) = -38$

35.   Adding: $(-5) + (-10) + (-7) = -22$

37.   Adding: $(-21) + 40 + (-23) + 5 = -44 + 45 = 1$

39.   Subtracting: $14 - (-8) = 14 + 8 = 22$

41.   Subtracting: $4 - 9 - 15 = 4 + (-9) + (-15) = 4 + (-24) = -20$

43.   Simplifying: $5 - (-10 - 2) - 3 = 5 - (-12) - 3 = 5 + 12 + (-3) = 17 + (-3) = 14$

45.   Simplifying: $20 - [-(10 - 3) - 8] = 20 - (-7 - 8) - 7 = 20 - (-15) - 7 = 20 + 15 + (-7) = 35 + (-7) = 28$

47.   Multiplying: $4(-3) = -12$

49.   Multiplying: $(-1)(-3)(-1)(-4) = 12$

51.   Finding the quotient: $\frac{-9}{36} = -\frac{1}{4}$

53.   Simplifying using order of operations: $4 \cdot 5 + 3 = 20 + 3 = 23$

55.   Simplifying using order of operations: $2^3 - 4 \cdot 3^2 + 5^2 = 8 - 4 \cdot 9 + 25 = 8 - 36 + 25 = 33 - 36 = -3$

57.   Simplifying using order of operations: $20 + 8 \div 4 + 2 \cdot 5 = 20 + 2 + 10 = 32$

59.   Simplifying using order of operations: $-4(-5) + 10 = 20 + 10 = 30$

61.   Simplifying using order of operations:
$$3(4 - 7)^2 - 5(3 - 8)^2 = 3(-3)^2 - 5(-5)^2 = 3 \cdot 9 - 5 \cdot 25 = 27 - 125 = 27 + (-125) = -98$$

63.   Simplifying using order of operations: $\dfrac{4(-3)}{-6} = \dfrac{-12}{-6} = 2$

65.   Simplifying using order of operations: $\dfrac{15 - 10}{6 - 6} = \dfrac{5}{0} =$ undefined

67.   Simplifying using order of operations: $\dfrac{2(-7) + (-11)(-4)}{7 - (-3)} = \dfrac{-14 + 44}{7 + 3} = \dfrac{30}{10} = 3$

69.   multiplicative identity property

71.   additive inverse property

73.   additive identity property

75.   distributive property

77.   Simplifying the expression: $4(7a) = (4 \cdot 7)a = 28a$

79.   Simplifying the expression: $\frac{4}{5}\left(\frac{5}{4}y\right) = \left(\frac{4}{5} \cdot \frac{5}{4}\right)y = 1y = y$

81.   Applying the distributive property: $3(2a - 4) = 3 \cdot 2a - 3 \cdot 4 = 6a - 12$

83.   Applying the distributive property: $-\frac{1}{2}(3x - 6) = -\frac{1}{2} \cdot 3x - \left(-\frac{1}{2}\right) \cdot 6 = -\frac{3}{2}x + 3$

85.   The whole numbers are: 0, 5   **87.**   The integers are: 0, 5, –3

89.   Factoring the number: $840 = 84 \cdot 10 = (21 \cdot 4) \cdot (2 \cdot 5) = (3 \cdot 7 \cdot 2 \cdot 2) \cdot (2 \cdot 5) = 2^3 \cdot 3 \cdot 5 \cdot 7$

91.   First factor the denominators to find the LCM:

$$70 = 2 \cdot 5 \cdot 7$$
$$84 = 2 \cdot 2 \cdot 3 \cdot 7$$
$$LCM = 2 \cdot 2 \cdot 3 \cdot 5 \cdot 7 = 420$$

Combining the fractions: $\frac{9}{70} + \frac{11}{84} = \frac{9 \cdot 6}{70 \cdot 6} + \frac{11 \cdot 5}{84 \cdot 5} = \frac{54}{420} + \frac{55}{420} = \frac{109}{420}$

93.   The pattern is to multiply by –3, so the next number is: $-270 \cdot (-3) = 810$

95.   The pattern is to add 2, so the next number is: $10 + 2 = 12$

97.   The pattern is to multiply by $-\frac{1}{2}$, so the next number is: $-\frac{1}{8} \cdot \left(-\frac{1}{2}\right) = \frac{1}{16}$

Chapter 1 Test

1.   Translating into symbols: $x + 3 = 8$ **2.**   Translating into symbols: $5y = 15$

3.   Simplifying using order of operations: $5^2 + 3(9 - 7) + 3^2 = 5^2 + 3(2) + 3^2 = 25 + 6 + 9 = 40$

4.   Simplifying using order of operations: $10 - 6 \div 3 + 2^3 = 10 - 6 \div 3 + 8 = 10 - 2 + 8 = 18 - 2 = 16$

5.   The opposite of –4 is 4, the reciprocal is $-\frac{1}{4}$, and the absolute value is $\left|-4\right| = 4$.

6.   The opposite of $\frac{3}{4}$ is $-\frac{3}{4}$, the reciprocal is $\frac{4}{3}$, and the absolute value is $\left|\frac{3}{4}\right| = \frac{3}{4}$.

7.   Adding: $3 + (-7) = -4$

8.   Adding: $\left|-9 + (-6)\right| + \left|-3 + 5\right| = \left|-15\right| + \left|2\right| = 15 + 2 = 17$

9.   Subtracting: $-4 - 8 = -4 + (-8) = -12$

10.   Subtracting: $9 - (7 - 2) - 4 = 9 - 5 - 4 = 9 + (-5) + (-4) = 9 + (-9) = 0$

11.   c (associative property of addition)   **12.**   e (distributive property)

13.   d (associative property of multiplication)   **14.**   a (commutative property of addition)

15.   Multiplying: $-3(7) = -21$ **16.**   Multiplying: $-4(8)(-2) = 64$

17.   Multiplying: $8\left(-\frac{1}{4}\right) = \frac{8}{1} \cdot \left(-\frac{1}{4}\right) = -\frac{8}{4} = -2$ **18.**   Multiplying: $\left(-\frac{2}{3}\right)^3 = \left(-\frac{2}{3}\right) \cdot \left(-\frac{2}{3}\right) \cdot \left(-\frac{2}{3}\right) = -\frac{8}{27}$

19.   Simplifying using order of operations: $-3(-4) - 8 = 12 - 8 = 4$

20.   Simplifying using order of operations: $5(-6)^2 - 3(-2)^3 = 5(36) - 3(-8) = 180 + 24 = 204$

21.   Simplifying using order of operations: $7 - 3(2 - 8) = 7 - 3(-6) = 7 + 18 = 25$

22   Simplifying using order of operations:

$$4 - 2\left[-3(-1 + 5) + 4(-3)\right] = 4 - 2\left[-3(4) + 4(-3)\right] = 4 - 2(-12 - 12) = 4 - 2(-24) = 4 + 48 = 52$$

23.   Simplifying using order of operations: $\dfrac{4(-5) - 2(7)}{-10 - 7} = \dfrac{-20 - 14}{-17} = \dfrac{-34}{-17} = 2$

24.   Simplifying using order of operations: $\dfrac{2(-3 - 1) + 4(-5 + 2)}{-3(2) - 4} = \dfrac{2(-4) + 4(-3)}{-6 - 4} = \dfrac{-8 - 12}{-10} = \dfrac{-20}{-10} = 2$

25.   Simplifying: $3 + (5 + 2x) = (3 + 5) + 2x = 8 + 2x$ **26.**   Simplifying: $-2(-5x) = [-2(-5)]x = 10x$

27.   Multiplying: $2(3x + 5) = 2 \cdot 3x + 2 \cdot 5 = 6x + 10$

28.   Multiplying: $-\frac{1}{2}(4x - 2) = -\frac{1}{2}(4x) - \left(-\frac{1}{2}\right)(2) = -2x + 1$

29.   The integers are 1 and –8.   **30.**   The rational numbers are 1, 1.5, $\frac{3}{4}$, and –8.

31.   The irrational numbers are $\sqrt{2}$.   **32.**   The real numbers are 1, 1.5, $\sqrt{2}$, $\frac{3}{4}$, and –8.

33.   Factoring the number: $592 = 4 \cdot 148 = (2 \cdot 2) \cdot (4 \cdot 37) = (2 \cdot 2) \cdot (2 \cdot 2 \cdot 37) = 2^4 \cdot 37$

34.   Factoring the number: $1340 = 10 \cdot 134 = (2 \cdot 5) \cdot (2 \cdot 67) = 2^2 \cdot 5 \cdot 67$

35. First factor the denominators to find the LCM:

$$15 = 3 \cdot 5$$
$$42 = 2 \cdot 3 \cdot 7$$
$$\text{LCM} = 2 \cdot 3 \cdot 5 \cdot 7 = 210$$

Combining the fractions: $\frac{5}{15} + \frac{11}{42} = \frac{5 \cdot 14}{15 \cdot 14} + \frac{11 \cdot 5}{42 \cdot 5} = \frac{70}{210} + \frac{55}{210} = \frac{125}{210} = \frac{5 \cdot 25}{5 \cdot 42} = \frac{25}{42}$

36. Combining the fractions: $\frac{5}{x} + \frac{3}{2} = \frac{5 \cdot 2}{x \cdot 2} + \frac{3 \cdot x}{2 \cdot x} = \frac{10}{2x} + \frac{3x}{2x} = \frac{10 + 3x}{2x}$

37. The expression is: $8 + (-3) = 5$
38. The expression is: $-24 - 2 = -24 + (-2) = -26$

39. The expression is: $-5(-4) = 20$
40. The expression is: $\frac{-24}{-2} = 12$

41. The pattern is to add 5, so the next term is $7 + 5 = 12$.

42. The pattern is to multiply by $-\frac{1}{2}$, so the next term is $-1\left(-\frac{1}{2}\right) = \frac{1}{2}$.

Chapter 2
Linear Equations and Inequalities

2.1 Simplifying Expressions

1. Simplifying the expression: $3x - 6x = (3 - 6)x = -3x$
3. Simplifying the expression: $-2a + a = (-2 + 1)a = -a$
5. Simplifying the expression: $7x + 3x + 2x = (7 + 3 + 2)x = 12x$
7. Simplifying the expression: $3a - 2a + 5a = (3 - 2 + 5)a = 6a$
9. Simplifying the expression: $4x - 3 + 2x = 4x + 2x - 3 = 6x - 3$
11. Simplifying the expression: $3a + 4a + 5 = 7a + 5$
13. Simplifying the expression: $2x - 3 + 3x - 2 = 2x + 3x - 3 - 2 = 5x - 5$
15. Simplifying the expression: $3a - 1 + a + 3 = 3a + a - 1 + 3 = 4a + 2$
17. Simplifying the expression: $-4x + 8 - 5x - 10 = -4x - 5x + 8 - 10 = -9x - 2$
19. Simplifying the expression: $7a + 3 + 2a + 3a = 7a + 2a + 3a + 3 = 12a + 3$
21. Simplifying the expression: $5(2x - 1) + 4 = 10x - 5 + 4 = 10x - 1$
23. Simplifying the expression: $7(3y + 2) - 8 = 21y + 14 - 8 = 21y + 6$
25. Simplifying the expression: $-3(2x - 1) + 5 = -6x + 3 + 5 = -6x + 8$
27. Simplifying the expression: $5 - 2(a + 1) = 5 - 2a - 2 = -2a - 2 + 5 = -2a + 3$
29. Simplifying the expression: $6 - 4(x - 5) = 6 - 4x + 20 = -4x + 20 + 6 = -4x + 26$
31. Simplifying the expression: $-9 - 4(2 - y) + 1 = -9 - 8 + 4y + 1 = 4y + 1 - 9 - 8 = 4y - 16$
33. Simplifying the expression: $-6 + 2(2 - 3x) + 1 = -6 + 4 - 6x + 1 = -6x - 6 + 4 + 1 = -6x - 1$
35. Simplifying the expression: $(4x - 7) - (2x + 5) = 4x - 7 - 2x - 5 = 4x - 2x - 7 - 5 = 2x - 12$
37. Simplifying the expression: $8(2a + 4) - (6a - 1) = 16a + 32 - 6a + 1 = 16a - 6a + 32 + 1 = 10a + 33$
39. Simplifying the expression: $3(x - 2) + (x - 3) = 3x - 6 + x - 3 = 3x + x - 6 - 3 = 4x - 9$
41. Simplifying the expression: $4(2y - 8) - (y + 7) = 8y - 32 - y - 7 = 8y - y - 32 - 7 = 7y - 39$
43. Simplifying the expression: $-9(2x + 1) - (x + 5) = -18x - 9 - x - 5 = -18x - x - 9 - 5 = -19x - 14$
45. Evaluating when $x = 2$: $3x - 1 = 3(2) - 1 = 6 - 1 = 5$
47. Evaluating when $x = 2$: $-2x - 5 = -2(2) - 5 = -4 - 5 = -9$
49. Evaluating when $x = 2$: $x^2 - 8x + 16 = (2)^2 - 8(2) + 16 = 4 - 16 + 16 = 4$
51. Evaluating when $x = 2$: $(x - 4)^2 = (2 - 4)^2 = (-2)^2 = 4$
53. Evaluating when $x = -5$: $7x - 4 - x - 3 = 7(-5) - 4 - (-5) - 3 = -35 - 4 + 5 - 3 = -42 + 5 = -37$
    Now simplifying the expression: $7x - 4 - x - 3 = 7x - x - 4 - 3 = 6x - 7$
    Evaluating when $x = -5$: $6x - 7 = 6(-5) - 7 = -30 - 7 = -37$
    Note that the two values are the same.
55. Evaluating when $x = -5$: $5(2x + 1) + 4 = 5[2(-5) + 1] + 4 = 5(-10 + 1) + 4 = 5(-9) + 4 = -45 + 4 = -41$
    Now simplifying the expression: $5(2x + 1) + 4 = 10x + 5 + 4 = 10x + 9$
    Evaluating when $x = -5$: $10x + 9 = 10(-5) + 9 = -50 + 9 = -41$
    Note that the two values are the same.

57.   Evaluating when $x = -3$ and $y = 5$: $x^2 - 2xy + y^2 = (-3)^2 - 2(-3)(5) + (5)^2 = 9 + 30 + 25 = 64$

59.   Evaluating when $x = -3$ and $y = 5$: $(x - y)^2 = (-3 - 5)^2 = (-8)^2 = 64$

61.   Evaluating when $x = -3$ and $y = 5$: $x^2 + 6xy + 9y^2 = (-3)^2 + 6(-3)(5) + 9(5)^2 = 9 - 90 + 225 = 144$

63.   Evaluating when $x = -3$ and $y = 5$: $(x + 3y)^2 = [-3 + 3(5)]^2 = (-3 + 15)^2 = (12)^2 = 144$

65.   Evaluating when $x = \frac{1}{2}$: $12x - 3 = 12\left(\frac{1}{2}\right) - 3 = 6 - 3 = 3$

67.   Evaluating when $x = \frac{1}{4}$: $12x - 3 = 12\left(\frac{1}{4}\right) - 3 = 3 - 3 = 0$

69.   Evaluating when $x = \frac{3}{2}$: $12x - 3 = 12\left(\frac{3}{2}\right) - 3 = 18 - 3 = 15$

71.   Evaluating when $x = \frac{3}{4}$: $12x - 3 = 12\left(\frac{3}{4}\right) - 3 = 9 - 3 = 6$

73.   **a.**   Substituting the values for n:

n	1	2	3	4
$3n$	3	6	9	12

   b.   Substituting the values for n:

n	1	2	3	4
n^3	1	8	27	64

75.   Substituting $n = 1, 2, 3, 4$:
$$n = 1: \quad 3(1) - 2 = 3 - 2 = 1$$
$$n = 2: \quad 3(2) - 2 = 6 - 2 = 4$$
$$n = 3: \quad 3(3) - 2 = 9 - 2 = 7$$
$$n = 4: \quad 3(4) - 2 = 12 - 2 = 10$$
The sequence is 1, 4, 7, 10, ..., which is an arithmetic sequence.

77.   Substituting $n = 1, 2, 3, 4$:
$$n = 1: \quad (1)^2 - 2(1) + 1 = 1 - 2 + 1 = 0$$
$$n = 2: \quad (2)^2 - 2(2) + 1 = 4 - 4 + 1 = 1$$
$$n = 3: \quad (3)^2 - 2(3) + 1 = 9 - 6 + 1 = 4$$
$$n = 4: \quad (4)^2 - 2(4) + 1 = 16 - 8 + 1 = 9$$
The sequence is 0, 1, 4, 9, ..., which is a sequence of squares.

79.   Translating into an algebraic expression: $x + 5$. Evaluating when $x = -2$: $x + 5 = (-2) + 5 = 3$

81.   Translating into an algebraic expression: $x - 5$. Evaluating when $x = -2$: $x - 5 = (-2) - 5 = -7$

83.   Translating into an algebraic expression: $2(x + 10)$. Evaluating when $x = -2$: $2(x + 10) = 2(-2 + 10) = 2(8) = 16$

85.   Translating into an algebraic expression: $\dfrac{10}{x}$. Evaluating when $x = -2$: $\dfrac{10}{x} = \dfrac{10}{-2} = -5$

87.   Translating into an algebraic expression, and simplifying: $[3x + (-2)] - 5 = 3x - 2 - 5 = 3x - 7$
Evaluating when $x = -2$: $3x - 7 = 3(-2) - 7 = -6 - 7 = -13$

89.   **a.**   Substituting $x = 8{,}000$: $-0.0035(8000) + 70 = 42°F$
   b.   Substituting $x = 12{,}000$: $-0.0035(12000) + 70 = 28°F$
   c.   Substituting $x = 24{,}000$: $-0.0035(24000) + 70 = -14°F$

91.   **a.**   Substituting $t = 10$: $35 + 0.25(10) = \$37.50$ **b.**   Substituting $t = 20$: $35 + 0.25(20) = \$40.00$
   c.   Substituting $t = 30$: $35 + 0.25(30) = \$42.50$

93.   Simplifying the expression: $G - 0.21G - 0.08G = 0.71G$. Substituting $G = \$1{,}250$: $0.71(\$1{,}250) = \$887.50$

95.   Subtracting: $-3 - \frac{1}{2} = \frac{-3}{1} - \frac{1}{2} = \frac{-3 \cdot 2}{1 \cdot 2} - \frac{1}{2} = \frac{-6}{2} - \frac{1}{2} = -\frac{7}{2}$

97.   Adding: $\frac{4}{5} + \frac{1}{10} + \frac{3}{8} = \frac{4 \cdot 8}{5 \cdot 8} + \frac{1 \cdot 4}{10 \cdot 4} + \frac{3 \cdot 5}{8 \cdot 5} = \frac{32}{40} + \frac{4}{40} + \frac{15}{40} = \frac{51}{40}$

2.2 Addition Property of Equality

1. Solving the equation:
$$x - 3 = 8$$
$$x - 3 + 3 = 8 + 3$$
$$x = 11$$

3. Solving the equation:
$$x + 2 = 6$$
$$x + 2 + (-2) = 6 + (-2)$$
$$x = 4$$

5. Solving the equation:
$$a + \tfrac{1}{2} = -\tfrac{1}{4}$$
$$a + \tfrac{1}{2} + \left(-\tfrac{1}{2}\right) = -\tfrac{1}{4} + \left(-\tfrac{1}{2}\right)$$
$$a = -\tfrac{1}{4} + \left(-\tfrac{2}{4}\right)$$
$$a = -\tfrac{3}{4}$$

7. Solving the equation:

$$x + 2.3 = -3.5$$
$$x + 2.3 + (-2.3) = -3.5 + (-2.3)$$
$$x = -5.8$$

9. Solving the equation:

$$y + 11 = -6$$
$$y + 11 + (-11) = -6 + (-11)$$
$$y = -17$$

11. Solving the equation:
$$x - \tfrac{5}{8} = -\tfrac{3}{4}$$
$$x - \tfrac{5}{8} + \tfrac{5}{8} = -\tfrac{3}{4} + \tfrac{5}{8}$$
$$x = -\tfrac{6}{8} + \tfrac{5}{8}$$
$$x = -\tfrac{1}{8}$$

13. Solving the equation:
$$m - 6 = -10$$
$$m - 6 + 6 = -10 + 6$$
$$m = -4$$

15. Solving the equation:
$$6.9 + x = 3.3$$
$$-6.9 + 6.9 + x = -6.9 + 3.3$$
$$x = -3.6$$

17. Solving the equation:
$$5 = a + 4$$
$$5 + (-4) = a + 4 + (-4)$$
$$a = 1$$

18. Solving the equation:
$$12 = a - 3$$
$$12 + 3 = a - 3 + 3$$
$$a = 15$$

19. Solving the equation:
$$-\tfrac{5}{9} = x - \tfrac{2}{5}$$
$$-\tfrac{5}{9} + \tfrac{2}{5} = x - \tfrac{2}{5} + \tfrac{2}{5}$$
$$-\tfrac{25}{45} + \tfrac{18}{45} = x$$
$$x = -\tfrac{7}{45}$$

21. Solving the equation:
$$-\tfrac{5}{9} = x - \tfrac{2}{5}$$
$$-\tfrac{5}{9} + \tfrac{2}{5} = x - \tfrac{2}{5} + \tfrac{2}{5}$$
$$-\tfrac{25}{45} + \tfrac{18}{45} = x$$
$$x = -\tfrac{7}{45}$$

23. Solving the equation:
$$8a - \tfrac{1}{2} - 7a = \tfrac{3}{4} + \tfrac{1}{8}$$
$$a - \tfrac{1}{2} = \tfrac{6}{8} + \tfrac{1}{8}$$
$$a - \tfrac{1}{2} = \tfrac{7}{8}$$
$$a - \tfrac{1}{2} + \tfrac{1}{2} = \tfrac{7}{8} + \tfrac{1}{2}$$
$$a = \tfrac{7}{8} + \tfrac{4}{8}$$
$$a = \tfrac{11}{8}$$

25. Solving the equation:

$$-3 - 4x + 5x = 18$$
$$-3 + x = 18$$
$$3 - 3 + x = 3 + 18$$
$$x = 21$$

27. Solving the equation:
$$-11x + 2 + 10x + 2x = 9$$
$$x + 2 = 9$$
$$x + 2 + (-2) = 9 + (-2)$$
$$x = 7$$

29. Solving the equation:
$$-2.5 + 4.8 = 8x - 1.2 - 7x$$
$$2.3 = x - 1.2$$
$$2.3 + 1.2 = x - 1.2 + 1.2$$
$$x = 3.5$$

31. Solving the equation:
$$2y - 10 + 3y - 4y = 18 - 6$$
$$y - 10 = 12$$
$$y - 10 + 10 = 12 + 10$$
$$y = 22$$

33. Solving the equation:
$$2(x + 3) - x = 4$$
$$2x + 6 - x = 4$$
$$x + 6 = 4$$
$$x + 6 + (-6) = 4 + (-6)$$
$$x = -2$$

35. Solving the equation:
$$-3(x-4)+4x = 3-7$$
$$-3x+12+4x = -4$$
$$x+12 = -4$$
$$x+12+(-12) = -4+(-12)$$
$$x = -16$$

37. Solving the equation:
$$5(2a+1)-9a = 8-6$$
$$10a+5-9a = 2$$
$$a+5 = 2$$
$$a+5+(-5) = 2+(-5)$$
$$a = -3$$

39. Solving the equation:
$$-(x+3)+2x-1 = 6$$
$$-x-3+2x-1 = 6$$
$$x-4 = 6$$
$$x-4+4 = 6+4$$
$$x = 10$$

41. Solving the equation:
$$4y-3(y-6)+2 = 8$$
$$4y-3y+18+2 = 8$$
$$y+20 = 8$$
$$y+20+(-20) = 8+(-20)$$
$$y = -12$$

43. Solving the equation:
$$-3(2m-9)+7(m-4) = 12-9$$
$$-6m+27+7m-28 = 3$$
$$m-1 = 3$$
$$m-1+1 = 3+1$$
$$m = 4$$

45. Solving the equation:
$$4x = 3x+2$$
$$4x+(-3x) = 3x+(-3x)+2$$
$$x = 2$$

47. Solving the equation:
$$8a = 7a-5$$
$$8a+(-7a) = 7a+(-7a)-5$$
$$a = -5$$

49. Solving the equation:
$$2x = 3x+1$$
$$(-2x)+2x = (-2x)+3x+1$$
$$0 = x+1$$
$$0+(-1) = x+1+(-1)$$
$$x = -1$$

51. Solving the equation:
$$3y+4 = 2y+1$$
$$3y+(-2y)+4 = 2y+(-2y)+1$$
$$y+4 = 1$$
$$y+4+(-4) = 1+(-4)$$
$$y = -3$$

53. Solving the equation:
$$2m-3 = m+5$$
$$2m+(-m)-3 = m+(-m)+5$$
$$m-3 = 5$$
$$m-3+3 = 5+3$$
$$m = 8$$

55. Solving the equation:
$$4x-7 = 5x+1$$
$$4x+(-4x)-7 = 5x+(-4x)+1$$
$$-7 = x+1$$
$$-7+(-1) = x+1+(-1)$$
$$x = -8$$

57. Solving the equation:
$$5x-\tfrac{2}{3} = 4x+\tfrac{4}{3}$$
$$5x+(-4x)-\tfrac{2}{3} = 4x+(-4x)+\tfrac{4}{3}$$
$$x-\tfrac{2}{3} = \tfrac{4}{3}$$
$$x-\tfrac{2}{3}+\tfrac{2}{3} = \tfrac{4}{3}+\tfrac{2}{3}$$
$$x = \tfrac{6}{3} = 2$$

59. Solving the equation:
$$8a-7.1 = 7a+3.9$$
$$8a+(-7a)-7.1 = 7a+(-7a)+3.9$$
$$a-7.1 = 3.9$$
$$a-7.1+7.1 = 3.9+7.1$$
$$a = 11$$

61. **a.** Solving for R:
$$T+R+A = 100$$
$$88+R+6 = 100$$
$$94+R = 100$$
$$R = 6\%$$

  b. Solving for R:
$$T+R+A = 100$$
$$0+R+95 = 100$$
$$95+R = 100$$
$$R = 5\%$$

  c. Solving for A:
$$T+R+A = 100$$
$$0+98+A = 100$$
$$98+A = 100$$
$$A = 2\%$$

  d. Solving for R:
$$T+R+A = 100$$
$$0+R+25 = 100$$
$$25+R = 100$$
$$R = 75\%$$

63. Solving for x:
$$x + 55 + 55 = 180$$
$$x + 110 = 180$$
$$x = 70°$$

65. Applying the associative property: $3(6x) = (3 \cdot 6)x = 18x$

67. Applying the associative property: $\frac{1}{5}(5x) = \left(\frac{1}{5} \cdot 5\right)x = 1x = x$

69. Applying the associative property: $8\left(\frac{1}{8}y\right) = \left(8 \cdot \frac{1}{8}\right)y = 1y = y$

71. Applying the associative property: $-2\left(-\frac{1}{2}x\right) = \left[-2 \cdot \left(-\frac{1}{2}\right)\right]x = 1x = x$

73. Applying the associative property: $-\frac{4}{3}\left(-\frac{3}{4}a\right) = \left[-\frac{4}{3} \cdot \left(-\frac{3}{4}\right)\right]a = 1a = a$

2.3 Multiplication Property of Equality

1. Solving the equation:
$$5x = 10$$
$$\tfrac{1}{5}(5x) = \tfrac{1}{5}(10)$$
$$x = 2$$

3. Solving the equation:
$$7a = 28$$
$$\tfrac{1}{7}(7a) = \tfrac{1}{7}(28)$$
$$a = 4$$

5. Solving the equation:
$$-8x = 4$$
$$-\tfrac{1}{8}(-8x) = -\tfrac{1}{8}(4)$$
$$x = -\tfrac{1}{2}$$

7. Solving the equation:
$$8m = -16$$
$$\tfrac{1}{8}(8m) = \tfrac{1}{8}(-16)$$
$$m = -2$$

9. Solving the equation:
$$-3x = -9$$
$$-\tfrac{1}{3}(-3x) = -\tfrac{1}{3}(-9)$$
$$x = 3$$

11. Solving the equation:
$$-7y = -28$$
$$-\tfrac{1}{7}(-7y) = -\tfrac{1}{7}(-28)$$
$$y = 4$$

13. Solving the equation:
$$2x = 0$$
$$\tfrac{1}{2}(2x) = \tfrac{1}{2}(0)$$
$$x = 0$$

15. Solving the equation:
$$-5x = 0$$
$$-\tfrac{1}{5}(-5x) = -\tfrac{1}{5}(0)$$
$$x = 0$$

17. Solving the equation:
$$\frac{x}{3} = 2$$
$$3\left(\frac{x}{3}\right) = 3(2)$$
$$x = 6$$

19. Solving the equation:
$$-\frac{m}{5} = 10$$
$$-5\left(-\frac{m}{5}\right) = -5(10)$$
$$m = -50$$

21. Solving the equation:
$$-\frac{x}{2} = -\frac{3}{4}$$
$$-2\left(-\frac{x}{2}\right) = -2\left(-\frac{3}{4}\right)$$
$$x = \frac{3}{2}$$

23. Solving the equation:
$$\tfrac{2}{3}a = 8$$
$$\tfrac{3}{2}\left(\tfrac{2}{3}a\right) = \tfrac{3}{2}(8)$$
$$a = 12$$

25. Solving the equation:
$$-\tfrac{3}{5}x = \tfrac{9}{5}$$
$$-\tfrac{5}{3}\left(-\tfrac{3}{5}x\right) = -\tfrac{5}{3}\left(\tfrac{9}{5}\right)$$
$$x = -3$$

27. Solving the equation:
$$-\tfrac{5}{8}y = -20$$
$$-\tfrac{8}{5}\left(-\tfrac{5}{8}y\right) = -\tfrac{8}{5}(-20)$$
$$y = 32$$

29.   Simplifying and then solving the equation:
$$-4x - 2x + 3x = 24$$
$$-3x = 24$$
$$-\tfrac{1}{3}(-3x) = -\tfrac{1}{3}(24)$$
$$x = -8$$

31.   Simplifying and then solving the equation:
$$4x + 8x - 2x = 15 - 10$$
$$10x = 5$$
$$\tfrac{1}{10}(10x) = \tfrac{1}{10}(5)$$
$$x = \tfrac{1}{2}$$

33.   Simplifying and then solving the equation:
$$-3 - 5 = 3x + 5x - 10x$$
$$-8 = -2x$$
$$-\tfrac{1}{2}(-8) = -\tfrac{1}{2}(-2x)$$
$$x = 4$$

35.   Using Method 2 to eliminate fractions:
$$18 - 13 = \tfrac{1}{2}a + \tfrac{3}{4}a - \tfrac{5}{8}a$$
$$8(5) = 8\left(\tfrac{1}{2}a + \tfrac{3}{4}a - \tfrac{5}{8}a\right)$$
$$40 = 4a + 6a - 5a$$
$$40 = 5a$$
$$\tfrac{1}{5}(40) = \tfrac{1}{5}(5a)$$
$$a = 8$$

37.   Solving by multiplying both sides of the equation by -1:
$$-x = 4$$
$$-1(-x) = -1(4)$$
$$x = -4$$

39.   Solving by multiplying both sides of the equation by -1:
$$-x = -4$$
$$-1(-x) = -1(-4)$$
$$x = 4$$

41.   Solving by multiplying both sides of the equation by -1:
$$15 = -a$$
$$-1(15) = -1(-a)$$
$$a = -15$$

43.   Solving by multiplying both sides of the equation by -1:
$$-y = \tfrac{1}{2}$$
$$-1(-y) = -1\left(\tfrac{1}{2}\right)$$
$$y = -\tfrac{1}{2}$$

45.   Solving the equation:
$$3x - 2 = 7$$
$$3x - 2 + 2 = 7 + 2$$
$$3x = 9$$
$$\tfrac{1}{3}(3x) = \tfrac{1}{3}(9)$$
$$x = 3$$

47.   Solving the equation:
$$2a + 1 = 3$$
$$2a + 1 + (-1) = 3 + (-1)$$
$$2a = 2$$
$$\tfrac{1}{2}(2a) = \tfrac{1}{2}(2)$$
$$a = 1$$

49.   Using Method 2 to eliminate fractions:
$$\tfrac{1}{8} + \tfrac{1}{2}x = \tfrac{1}{4}$$
$$8\left(\tfrac{1}{8} + \tfrac{1}{2}x\right) = 8\left(\tfrac{1}{4}\right)$$
$$1 + 4x = 2$$
$$(-1) + 1 + 4x = (-1) + 2$$
$$4x = 1$$
$$\tfrac{1}{4}(4x) = \tfrac{1}{4}(1)$$
$$x = \tfrac{1}{4}$$

51.   Solving the equation:
$$6x = 2x - 12$$
$$6x + (-2x) = 2x + (-2x) - 12$$
$$4x = -12$$
$$\tfrac{1}{4}(4x) = \tfrac{1}{4}(-12)$$
$$x = -3$$

53.   Solving the equation:
$$2y = -4y + 18$$
$$2y + 4y = -4y + 4y + 18$$
$$6y = 18$$
$$\tfrac{1}{6}(6y) = \tfrac{1}{6}(18)$$
$$y = 3$$

55.   Solving the equation:
$$-7x = -3x - 8$$
$$-7x + 3x = -3x + 3x - 8$$
$$-4x = -8$$
$$-\tfrac{1}{4}(-4x) = -\tfrac{1}{4}(-8)$$
$$x = 2$$

57. Solving the equation:

$$8x+4=2x-5$$
$$8x+(-2x)+4=2x+(-2x)-5$$
$$6x+4=-5$$
$$6x+4+(-4)=-5+(-4)$$
$$6x=-9$$
$$\tfrac{1}{6}(6x)=\tfrac{1}{6}(-9)$$
$$x=-\tfrac{3}{2}$$

59. Using Method 2 to eliminate fractions:

$$x+\tfrac{1}{2}=\tfrac{1}{4}x-\tfrac{5}{8}$$
$$8\left(x+\tfrac{1}{2}\right)=8\left(\tfrac{1}{4}x-\tfrac{5}{8}\right)$$
$$8x+4=2x-5$$
$$8x+(-2x)+4=2x+(-2x)-5$$
$$6x+4=-5$$
$$6x+4+(-4)=-5+(-4)$$
$$6x=-9$$
$$\tfrac{1}{6}(6x)=\tfrac{1}{6}(-9)$$
$$x=-\tfrac{3}{2}$$

61. Solving the equation:

$$6m-3=m+2$$
$$6m+(-m)-3=m+(-m)+2$$
$$5m-3=2$$
$$5m-3+3=2+3$$
$$5m=5$$
$$\tfrac{1}{5}(5m)=\tfrac{1}{5}(5)$$
$$m=1$$

63. Using Method 2 to eliminate fractions:

$$\tfrac{1}{2}m-\tfrac{1}{4}=\tfrac{1}{12}m+\tfrac{1}{6}$$
$$12\left(\tfrac{1}{2}m-\tfrac{1}{4}\right)=12\left(\tfrac{1}{12}m+\tfrac{1}{6}\right)$$
$$6m-3=m+2$$
$$6m+(-m)-3=m+(-m)+2$$
$$5m-3=2$$
$$5m-3+3=2+3$$
$$5m=5$$
$$\tfrac{1}{5}(5m)=\tfrac{1}{5}(5)$$
$$m=1$$

65. Solving the equation:

$$9y+2=6y-4$$
$$9y+(-6y)+2=6y+(-6y)-4$$
$$3y+2=-4$$
$$3y+2+(-2)=-4+(-2)$$
$$3y=-6$$
$$\tfrac{1}{3}(3y)=\tfrac{1}{3}(-6)$$
$$y=-2$$

67. Using Method 2 to eliminate fractions:

$$\tfrac{3}{2}y+\tfrac{1}{3}=y-\tfrac{2}{3}$$
$$6\left(\tfrac{3}{2}y+\tfrac{1}{3}\right)=6\left(y-\tfrac{2}{3}\right)$$
$$9y+2=6y-4$$
$$9y+(-6y)+2=6y+(-6y)-4$$
$$3y+2=-4$$
$$3y+2+(-2)=-4+(-2)$$
$$3y=-6$$
$$\tfrac{1}{3}(3y)=\tfrac{1}{3}(-6)$$
$$y=-2$$

69. Solving the equation:
$$7.5x=1500$$
$$\frac{7.5x}{7.5}=\frac{1500}{7.5}$$
$$x=200$$
The break-even point is 200 tickets.

71. Solving the equation:
$$1+3(2)+3x=13$$
$$7+3x=13$$
$$3x=6$$
$$x=2$$
Laura made 2 three-point shots.

73. a. Solving the proportion:
$$\frac{AX}{12}=\frac{15}{20}$$
$$20(AX)=180$$
$$AX=9$$

b. Solving the proportion:
$$\frac{12}{AB}=\frac{8}{10}$$
$$8(AB)=120$$
$$AB=15$$

c. Since $AX=8$ and $XB=4$, then $AB=12$. Also since $YC=6, AY=AC-6$. Solving the proportion:
$$\frac{AX}{AB}=\frac{AY}{AC}$$
$$\frac{8}{12}=\frac{AC-6}{AC}$$
$$8(AC)=12(AC)-72$$
$$-4(AC)=-72$$
$$AC=18$$

75.   Solving the equation:
$$3x + 2 = 19$$
$$3x + 2 - 2 = 19 - 2$$
$$3x = 17$$
$$x = \tfrac{17}{3}$$

77.   Solving the equation:
$$2(x + 10) = 40$$
$$2x + 20 = 40$$
$$2x + 20 - 20 = 40 - 20$$
$$2x = 20$$
$$x = 10$$

79.   Using the distributive property and combining like terms: $5(2x - 8) - 3 = 10x - 40 - 3 = 10x - 43$

81.   Using the distributive property and combining like terms:
$$-2(3x + 5) + 3(x - 1) = -6x - 10 + 3x - 3 = -6x + 3x - 10 - 3 = -3x - 13$$

83.   Using the distributive property and combining like terms: $7 - 3(2y + 1) = 7 - 6y - 3 = -6y + 7 - 3 = -6y + 4$

85.   Using the distributive property and combining like terms: $4x - (9x - 3) + 4 = 4x - 9x + 3 + 4 = -5x + 7$

2.4 Solving Linear Equations

1.   Solving the equation:
$$2(x + 3) = 12$$
$$2x + 6 = 12$$
$$2x + 6 + (-6) = 12 + (-6)$$
$$2x = 6$$
$$\tfrac{1}{2}(2x) = \tfrac{1}{2}(6)$$
$$x = 3$$

3.   Solving the equation:
$$6(x - 1) = -18$$
$$6x - 6 = -18$$
$$6x - 6 + 6 = -18 + 6$$
$$6x = -12$$
$$\tfrac{1}{6}(6x) = \tfrac{1}{6}(-12)$$
$$x = -2$$

5.   Solving the equation:
$$2(4a + 1) = -6$$
$$8a + 2 = -6$$
$$8a + 2 + (-2) = -6 + (-2)$$
$$8a = -8$$
$$\tfrac{1}{8}(8a) = \tfrac{1}{8}(-8)$$
$$a = -1$$

7.   Solving the equation:
$$14 = 2(5x - 3)$$
$$14 = 10x - 6$$
$$14 + 6 = 10x - 6 + 6$$
$$20 = 10x$$
$$\tfrac{1}{10}(20) = \tfrac{1}{10}(10x)$$
$$x = 2$$

9.   Solving the equation:
$$-2(3y + 5) = 14$$
$$-6y - 10 = 14$$
$$-6y - 10 + 10 = 14 + 10$$
$$-6y = 24$$
$$-\tfrac{1}{6}(-6y) = -\tfrac{1}{6}(24)$$
$$y = -4$$

11.   Solving the equation:
$$-5(2a + 4) = 0$$
$$-10a - 20 = 0$$
$$-10a - 20 + 20 = 0 + 20$$
$$-10a = 20$$
$$-\tfrac{1}{10}(-10a) = -\tfrac{1}{10}(20)$$
$$a = -2$$

13.   Solving the equation:
$$1 = \tfrac{1}{2}(4x + 2)$$
$$1 = 2x + 1$$
$$1 + (-1) = 2x + 1 + (-1)$$
$$0 = 2x$$
$$\tfrac{1}{2}(0) = \tfrac{1}{2}(2x)$$
$$x = 0$$

15.   Solving the equation:
$$3(t - 4) + 5 = -4$$
$$3t - 12 + 5 = -4$$
$$3t - 7 = -4$$
$$3t - 7 + 7 = -4 + 7$$
$$3t = 3$$
$$\tfrac{1}{3}(3t) = \tfrac{1}{3}(3)$$
$$t = 1$$

17.   Solving the equation:
$$4(2y + 1) - 7 = 1$$
$$8y + 4 - 7 = 1$$
$$8y - 3 = 1$$
$$8y - 3 + 3 = 1 + 3$$
$$8y = 4$$
$$\tfrac{1}{8}(8y) = \tfrac{1}{8}(4)$$
$$y = \tfrac{1}{2}$$

19.   Solving the equation:
$$\tfrac{1}{2}(x - 3) = \tfrac{1}{4}(x + 1)$$
$$\tfrac{1}{2}x - \tfrac{3}{2} = \tfrac{1}{4}x + \tfrac{1}{4}$$
$$4\left(\tfrac{1}{2}x - \tfrac{3}{2}\right) = 4\left(\tfrac{1}{4}x + \tfrac{1}{4}\right)$$
$$2x - 6 = x + 1$$
$$2x + (-x) - 6 = x + (-x) + 1$$
$$x - 6 = 1$$
$$x - 6 + 6 = 1 + 6$$
$$x = 7$$

21. Solving the equation:
$$-0.7(2x-7) = 0.3(11-4x)$$
$$-1.4x + 4.9 = 3.3 - 1.2x$$
$$-1.4x + 1.2x + 4.9 = 3.3 - 1.2x + 1.2x$$
$$-0.2x + 4.9 = 3.3$$
$$-0.2x + 4.9 + (-4.9) = 3.3 + (-4.9)$$
$$-0.2x = -1.6$$
$$\frac{-0.2x}{-0.2} = \frac{-1.6}{-0.2}$$
$$x = 8$$

23. Solving the equation:
$$-2(3y+1) = 3(1-6y) - 9$$
$$-6y - 2 = 3 - 18y - 9$$
$$-6y - 2 = -18y - 6$$
$$-6y + 18y - 2 = -18y + 18y - 6$$
$$12y - 2 = -6$$
$$12y - 2 + 2 = -6 + 2$$
$$12y = -4$$
$$\tfrac{1}{12}(12y) = \tfrac{1}{12}(-4)$$
$$y = -\tfrac{1}{3}$$

25. Solving the equation:
$$\tfrac{3}{4}(8x-4) + 3 = \tfrac{2}{5}(5x+10) - 1$$
$$6x - 3 + 3 = 2x + 4 - 1$$
$$6x = 2x + 3$$
$$6x + (-2x) = 2x + (-2x) + 3$$
$$4x = 3$$
$$\tfrac{1}{4}(4x) = \tfrac{1}{4}(3)$$
$$x = \tfrac{3}{4}$$

27. Solving the equation:
$$0.06x + 0.08(100 - x) = 6.5$$
$$0.06x + 8 - 0.08x = 6.5$$
$$-0.02x + 8 = 6.5$$
$$-0.02x + 8 + (-8) = 6.5 + (-8)$$
$$-0.02x = -1.5$$
$$\frac{-0.02x}{-0.02} = \frac{-1.5}{-0.02}$$
$$x = 75$$

29. Solving the equation:
$$6 - 5(2a-3) = 1$$
$$6 - 10a + 15 = 1$$
$$-10a + 21 = 1$$
$$-10a + 21 + (-21) = 1 + (-21)$$
$$-10a = -20$$
$$-\tfrac{1}{10}(-10a) = -\tfrac{1}{10}(-20)$$
$$a = 2$$

31. Solving the equation:
$$0.2x - 0.5 = 0.5 - 0.2(2x - 13)$$
$$0.2x - 0.5 = 0.5 - 0.4x + 2.6$$
$$0.2x - 0.5 = -0.4x + 3.1$$
$$0.2x + 0.4x - 0.5 = -0.4x + 0.4x + 3.1$$
$$0.6x - 0.5 = 3.1$$
$$0.6x - 0.5 + 0.5 = 3.1 + 0.5$$
$$0.6x = 3.6$$
$$\frac{0.6x}{0.6} = \frac{3.6}{0.6}$$
$$x = 6$$

33. Solving the equation:
$$2(t-3) + 3(t-2) = 28$$
$$2t - 6 + 3t - 6 = 28$$
$$5t - 12 = 28$$
$$5t - 12 + 12 = 28 + 12$$
$$5t = 40$$
$$\tfrac{1}{5}(5t) = \tfrac{1}{5}(40)$$
$$t = 8$$

35. Solving the equation:
$$5(x-2) - (3x+4) = 3(6x-8) + 10$$
$$5x - 10 - 3x - 4 = 18x - 24 + 10$$
$$2x - 14 = 18x - 14$$
$$2x + (-18x) - 14 = 18x + (-18x) - 14$$
$$-16x - 14 = -14$$
$$-16x - 14 + 14 = -14 + 14$$
$$-16x = 0$$
$$-\tfrac{1}{16}(-16x) = -\tfrac{1}{16}(0)$$
$$x = 0$$

37. Solving the equation:
$$2(5x-3) - (2x-4) = 5 - (6x+1)$$
$$10x - 6 - 2x + 4 = 5 - 6x - 1$$
$$8x - 2 = -6x + 4$$
$$8x + 6x - 2 = -6x + 6x + 4$$
$$14x - 2 = 4$$
$$14x - 2 + 2 = 4 + 2$$
$$14x = 6$$
$$\tfrac{1}{14}(14x) = \tfrac{1}{14}(6)$$
$$x = \tfrac{3}{7}$$

39. Solving the equation:
$$-(3x+1) - (4x-7) = 4 - (3x+2)$$
$$-3x - 1 - 4x + 7 = 4 - 3x - 2$$
$$-7x + 6 = -3x + 2$$
$$-7x + 3x + 6 = -3x + 3x + 2$$
$$-4x + 6 = 2$$
$$-4x + 6 + (-6) = 2 + (-6)$$
$$-4x = -4$$
$$-\tfrac{1}{4}(-4x) = -\tfrac{1}{4}(-4)$$
$$x = 1$$

41. Multiplying the fractions: $\tfrac{1}{2}(3) = \tfrac{1}{2} \cdot \tfrac{3}{1} = \tfrac{3}{2}$

43. Multiplying the fractions: $\tfrac{2}{3}(6) = \tfrac{2}{3} \cdot \tfrac{6}{1} = \tfrac{12}{3} = 4$

45. Multiplying the fractions: $\frac{5}{9} \cdot \frac{9}{5} = \frac{45}{45} = 1$

47. Applying the distributive property: $2(3x-5) = 2 \cdot 3x - 2 \cdot 5 = 6x - 10$

49. Applying the distributive property: $\frac{1}{2}(3x+6) = \frac{1}{2} \cdot 3x + \frac{1}{2} \cdot 6 = \frac{3}{2}x + 3$

51. Applying the distributive property: $\frac{1}{3}(-3x+6) = \frac{1}{3} \cdot (-3x) + \frac{1}{3} \cdot 6 = -x + 2$

2.5 Formulas

1. Using the perimeter formula:
$$P = 2l + 2w$$
$$300 = 2l + 2(50)$$
$$300 = 2l + 100$$
$$200 = 2l$$
$$l = 100$$
The length is 100 feet.

3. Substituting $x = 3$:
$$2(3) + 3y = 6$$
$$6 + 3y = 6$$
$$6 + (-6) + 3y = 6 + (-6)$$
$$3y = 0$$
$$y = 0$$

5. Substituting $x = 0$:
$$2(0) + 3y = 6$$
$$0 + 3y = 6$$
$$3y = 6$$
$$y = 2$$

7. Substituting $y = 2$:
$$2x - 5(2) = 20$$
$$2x - 10 = 20$$
$$2x - 10 + 10 = 20 + 10$$
$$2x = 30$$
$$x = 15$$

9. Substituting $y = 0$:
$$2x - 5(0) = 20$$
$$2x - 0 = 20$$
$$2x = 20$$
$$x = 10$$

11. Substituting $y = 7$:
$$7 = 2x - 1$$
$$7 + 1 = 2x - 1 + 1$$
$$2x = 8$$
$$x = 4$$

13. Substituting $y = 3$:
$$3 = 2x - 1$$
$$3 + 1 = 2x - 1 + 1$$
$$2x = 4$$
$$x = 2$$

15. Solving for l:
$$lw = A$$
$$\frac{lw}{w} = \frac{A}{w}$$
$$l = \frac{A}{w}$$

17. Solving for r:
$$rt = d$$
$$\frac{rt}{t} = \frac{d}{t}$$
$$r = \frac{d}{t}$$

19. Solving for h:
$$lwh = V$$
$$\frac{lwh}{lw} = \frac{V}{lw}$$
$$h = \frac{V}{lw}$$

21. Solving for P:
$$PV = nRT$$
$$\frac{PV}{V} = \frac{nRT}{V}$$
$$P = \frac{nRT}{V}$$

23. Solving for a:
$$a + b + c = P$$
$$a + b + c - b - c = P - b - c$$
$$a = P - b - c$$

25. Solving for x:
$$x - 3y = -1$$
$$x - 3y + 3y = -1 + 3y$$
$$x = 3y - 1$$

27. Solving for y:
$$-3x + y = 6$$
$$-3x + 3x + y = 6 + 3x$$
$$y = 3x + 6$$

29. Solving for y:
$$2x + 3y = 6$$
$$-2x + 2x + 3y = -2x + 6$$
$$3y = -2x + 6$$
$$\tfrac{1}{3}(3y) = \tfrac{1}{3}(-2x + 6)$$
$$y = -\tfrac{2}{3}x + 2$$

31. Solving for y:
$$6x + 3y = 12$$
$$-6x + 6x + 3y = -6x + 12$$
$$3y = -6x + 12$$
$$\tfrac{1}{3}(3y) = \tfrac{1}{3}(-6x + 12)$$
$$y = -2x + 4$$

33. Solving for y:
$$5x - 2y = 3$$
$$-5x + 5x - 2y = -5x + 3$$
$$-2y = -5x + 3$$
$$-\tfrac{1}{2}(-2y) = -\tfrac{1}{2}(-5x + 3)$$
$$y = \tfrac{5}{2}x - \tfrac{3}{2}$$

35. Solving for w:
$$2l + 2w = P$$
$$2l - 2l + 2w = P - 2l$$
$$2w = P - 2l$$
$$\frac{2w}{2} = \frac{P - 2l}{2}$$
$$w = \frac{P - 2l}{2}$$

37. Solving for v:
$$vt + 16t^2 = h$$
$$vt + 16t^2 - 16t^2 = h - 16t^2$$
$$vt = h - 16t^2$$
$$\frac{vt}{t} = \frac{h - 16t^2}{t}$$
$$v = \frac{h - 16t^2}{t}$$

39. Solving for h:
$$\pi r^2 + 2\pi rh = A$$
$$\pi r^2 - \pi r^2 + 2\pi rh = A - \pi r^2$$
$$2\pi rh = A - \pi r^2$$
$$\frac{2\pi rh}{2\pi r} = \frac{A - \pi r^2}{2\pi r}$$
$$h = \frac{A - \pi r^2}{2\pi r}$$

41. Solving for y:
$$\frac{x}{2} + \frac{y}{3} = 1$$
$$-\frac{x}{2} + \frac{x}{2} + \frac{y}{3} = -\frac{x}{2} + 1$$
$$\frac{y}{3} = -\frac{x}{2} + 1$$
$$3\left(\frac{y}{3}\right) = 3\left(-\frac{x}{2} + 1\right)$$
$$y = -\tfrac{3}{2}x + 3$$

43. Solving for y:
$$\frac{x}{7} - \frac{y}{3} = 1$$
$$-\frac{x}{7} + \frac{x}{7} - \frac{y}{3} = -\frac{x}{7} + 1$$
$$-\frac{y}{3} = -\frac{x}{7} + 1$$
$$-3\left(-\frac{y}{3}\right) = -3\left(-\frac{x}{7} + 1\right)$$
$$y = \tfrac{3}{7}x - 3$$

45. Solving for y:
$$-\tfrac{1}{4}x + \tfrac{1}{8}y = 1$$
$$-\tfrac{1}{4}x + \tfrac{1}{4}x + \tfrac{1}{8}y = 1 + \tfrac{1}{4}x$$
$$\tfrac{1}{8}y = \tfrac{1}{4}x + 1$$
$$8\left(\tfrac{1}{8}y\right) = 8\left(\tfrac{1}{4}x + 1\right)$$
$$y = 2x + 8$$

47. The complement of $30°$ is $90° - 30° = 60°$, and the supplement is $180° - 30° = 150°$.

49. The complement of $45°$ is $90° - 45° = 45°$, and the supplement is $180° - 45° = 135°$.

51. Translating into an equation and solving:
$$x = 0.25 \cdot 40$$
$$x = 10$$
The number 10 is 25% of 40.

53. Translating into an equation and solving:
$$x = 0.12 \cdot 2000$$
$$x = 240$$
The number 240 is 12% of 2000.

55. Translating into an equation and solving:
$$x \cdot 28 = 7$$
$$28x = 7$$
$$\tfrac{1}{28}(28x) = \tfrac{1}{28}(7)$$
$$x = 0.25 = 25\%$$
The number 7 is 25% of 28.

57. Translating into an equation and solving:
$$x \cdot 40 = 14$$
$$40x = 14$$
$$\tfrac{1}{40}(40x) = \tfrac{1}{40}(14)$$
$$x = 0.35 = 35\%$$
The number 14 is 35% of 40.

59.   Translating into an equation and solving:
$$0.50 \cdot x = 32$$
$$\frac{0.50x}{0.50} = \frac{32}{0.50}$$
$$x = 64$$
The number 32 is 50% of 64.

61.   Translating into an equation and solving:
$$0.12 \cdot x = 240$$
$$\frac{0.12x}{0.12} = \frac{240}{0.12}$$
$$x = 2000$$
The number 240 is 12% of 2000.

63.   Substituting $F = 212$: $C = \frac{5}{9}(212 - 32) = \frac{5}{9}(180) = 100°C$. This value agrees with the information in Table 1.

65.   Substituting $F = 68$: $C = \frac{5}{9}(68 - 32) = \frac{5}{9}(36) = 20°C$. This value agrees with the information in Table 1.

67.   Solving for C:
$$\frac{9}{5}C + 32 = F$$
$$\frac{9}{5}C + 32 - 32 = F - 32$$
$$\frac{9}{5}C = F - 32$$
$$\frac{5}{9}\left(\frac{9}{5}C\right) = \frac{5}{9}(F - 32)$$
$$C = \frac{5}{9}(F - 32)$$

69.   We need to find what percent of 150 is 90:
$$x \cdot 150 = 90$$
$$\frac{1}{150}(150x) = \frac{1}{150}(90)$$
$$x = 0.60 = 60\%$$
So 60% of the calories in one serving of vanilla ice cream are fat calories.

71.   We need to find what percent of 98 is 26:
$$x \cdot 98 = 26$$
$$\frac{1}{98}(98x) = \frac{1}{98}(26)$$
$$x \approx 0.265 = 26.5\%$$
So 26.5% of one serving of frozen yogurt are carbohydrates.

73.   Solving for r:
$$2\pi r = C$$
$$2 \cdot \frac{22}{7}r = 44$$
$$\frac{44}{7}r = 44$$
$$\frac{7}{44}\left(\frac{44}{7}r\right) = \frac{7}{44}(44)$$
$$r = 7 \text{ meters}$$

75.   Solving for r:
$$2\pi r = C$$
$$2 \cdot 3.14r = 9.42$$
$$6.28r = 9.42$$
$$\frac{6.28r}{6.28} = \frac{9.42}{6.28}$$
$$r = 1.5 \text{ inches}$$

77.   Solving for h:
$$\pi r^2 h = V$$
$$\frac{22}{7}\left(\frac{7}{22}\right)^2 h = 42$$
$$\frac{7}{22}h = 42$$
$$\frac{22}{7}\left(\frac{7}{22}h\right) = \frac{22}{7}(42)$$
$$h = 132 \text{ feet}$$

79.   Solving for h:
$$\pi r^2 h = V$$
$$3.14(3)^2 h = 6.28$$
$$28.26h = 6.28$$
$$\frac{28.26h}{28.26} = \frac{6.28}{28.26}$$
$$h = \frac{2}{9} \text{ centimeters}$$

81.   Completing the table:

Pitcher, Team	W	L	Saves	Blown Saves	Rolaids Points
John Franco, New York	0	8	38	8	82
Billy Wagner, Houston	4	3	30	5	82
Gregg Olson, Arizona	3	4	30	4	80
Bob Wickman, Milwaukee	6	9	25	7	55

83. **a.**   The percent of silver is: $\frac{12}{20} \cdot 100 = 60\%$ **b.**   The percent of copper is: $\frac{8}{20} \cdot 100 = 40\%$

85.   Completing the table:

Altitude (feet)	Temperature (°F)
0	56
9,000	24.5
15,000	3.5
21,000	−17.5
28,000	−42
40,000	−84

87.   The sum of 4 and 1 is 5.   **89.**   The difference of 6 and 2 is 4.

91.   An equivalent expression is: $2(6+3) = 2(9) = 18$ **93.**   An equivalent expression is: $2(5)+3 = 10+3 = 13$

2.6 Applications

1.   Let x represent the number. The equation is:

$$x + 5 = 13$$
$$x = 8$$

The number is 8.

3.   Let x represent the number. The equation is:

$$2x + 4 = 14$$
$$2x = 10$$
$$x = 5$$

The number is 5.

5.   Let x represent the number. The equation is:

$$5(x + 7) = 30$$
$$5x + 35 = 30$$
$$5x = -5$$
$$x = -1$$

The number is −1.

7.   Let x and $x + 2$ represent the two numbers. The equation is:

$$x + x + 2 = 8$$
$$2x + 2 = 8$$
$$2x = 6$$
$$x = 3$$
$$x + 2 = 5$$

The two numbers are 3 and 5.

9.   Let x and $3x - 4$ represent the two numbers. The equation is:

$$(x + 3x - 4) + 5 = 25$$
$$4x + 1 = 25$$
$$4x = 24$$
$$x = 6$$
$$3x - 4 = 3(6) - 4 = 14$$

The two numbers are 6 and 14.

11.   Completing the table:

	Five Years Ago	Now
Fred	$x+4-5 = x-1$	$x+4$
Barney	$x-5$	x

The equation is:

$$x - 1 + x - 5 = 48$$
$$2x - 6 = 48$$
$$2x = 54$$
$$x = 27$$
$$x + 4 = 31$$

Barney is 27 and Fred is 31.

13.   Completing the table:

	Now	Three Years from Now
Jack	$2x$	$2x+3$
Lacy	x	$x+3$

The equation is:

$$2x + 3 + x + 3 = 54$$
$$3x + 6 = 54$$
$$3x = 48$$
$$x = 16$$
$$2x = 32$$

Lacy is 16 and Jack is 32.

15.   Completing the table:

	Now	Two Years from Now
Pat	$x+20$	$x+20+2 = x+22$
Patrick	x	$x+2$

The equation is:
$$x+22 = 2(x+2)$$
$$x+22 = 2x+4$$
$$22 = x+4$$
$$x = 18$$
$$x+20 = 38$$
Patrick is 18 and Pat is 38.

17.   Let w represent the width and $w+5$ represent the length. The equation is:
$$2w+2(w+5) = 34$$
$$2w+2w+10 = 34$$
$$4w+10 = 34$$
$$4w = 24$$
$$w = 6$$
$$w+5 = 11$$
The length is 11 inches and the width is 6 inches.

19.   Let s represent the side of the square. The equation is:
$$4s = 48$$
$$s = 12$$
The length of one side is 12 meters.

21.   Let w represent the width and $2w-3$ represent the length. The equation is:
$$2w+2(2w-3) = 54$$
$$2w+4w-6 = 54$$
$$6w-6 = 54$$
$$6w = 60$$
$$w = 10$$
$$2w-3 = 2(10)-3 = 17$$
The length is 17 inches and the width is 10 inches.

23.   Completing the table:

	Nickels	Dimes
Number	x	$x+9$
Value (cents)	$5(x)$	$10(x+9)$

The equation is:
$$5(x)+10(x+9) = 210$$
$$5x+10x+90 = 210$$
$$15x+90 = 210$$
$$15x = 120$$
$$x = 8$$
$$x+9 = 17$$
Sue has 8 nickels and 17 dimes.

25.   Completing the table:

	Dimes	Quarters
Number	x	$2x$
Value (cents)	$10(x)$	$25(2x)$

The equation is:
$$10(x)+25(2x) = 900$$
$$10x+50x = 900$$
$$60x = 900$$
$$x = 15$$
$$2x = 30$$

You have 15 dimes and 30 quarters.

27.   Completing the table:

	Nickels	Dimes	Quarters
Number	x	$x+3$	$x+5$
Value (cents)	$5(x)$	$10(x+3)$	$25(x+5)$

The equation is:
$$5(x)+10(x+3)+25(x+5)=435$$
$$5x+10x+30+25x+125=435$$
$$40x+155=435$$
$$40x=280$$
$$x=7$$
$$x+3=10$$
$$x+5=12$$
Katie has 7 nickels, 10 dimes, and 12 quarters.

29.   The statement is: 4 is less than 10

31.   The statement is: 9 is greater than or equal to -5

33.   The correct statement is: $12 < 20$

35.   The correct statement is: $-8 < -6$

37.   Simplifying: $|8-3|-|5-2|=|5|-|3|=5-3=2$

39.   Simplifying: $15-|9-3(7-5)|=15-|9-3(2)|=15-|9-6|=15-|3|=15-3=12$

2.7 More Applications

1.   Completing the table:

	Dollars Invested at 8%	Dollars Invested at 9%
Number of	x	$x+2000$
Interest on	$0.08(x)$	$0.09(x+2000)$

The equation is:
$$0.08(x)+0.09(x+2000)=860$$
$$0.08x+0.09x+180=860$$
$$0.17x+180=860$$
$$0.17x=680$$
$$x=4000$$
$$x+2000=6000$$
You have $4,000 invested at 8% and $6,000 invested at 9%.

3.   Completing the table:

	Dollars Invested at 10%	Dollars Invested at 12%
Number of	x	$x+500$
Interest on	$0.10(x)$	$0.12(x+500)$

The equation is:
$$0.10(x)+0.12(x+500)=214$$
$$0.10x+0.12x+60=214$$
$$0.22x+60=214$$
$$0.22x=154$$
$$x=700$$
$$x+500=1200$$
Tyler has $700 invested at 10% and $1,200 invested at 12%.

5.   Completing the table:

	Dollars Invested at 8%	Dollars Invested at 9%	Dollars Invested at 10%
Number of	x	$2x$	$3x$
Interest on	$0.08(x)$	$0.09(2x)$	$0.10(3x)$

The equation is:
$$0.08(x) + 0.09(2x) + 0.10(3x) = 280$$
$$0.08x + 0.18x + 0.30x = 280$$
$$0.56x = 280$$
$$x = 500$$
$$2x = 1000$$
$$3x = 1500$$
She has $500 invested at 8%, $1,000 invested at 9%, and $1,500 invested at 10%.

7.   Let x represent the measure of the two equal angles, so $x + x = 2x$ represents the measure of the third angle. Since the sum of the three angles is 180°, the equation is:
$$x + x + 2x = 180°$$
$$4x = 180°$$
$$x = 45°$$
$$2x = 90°$$
The measures of the three angles are 45°, 45°, and 90°.

9.   Let x represent the measure of the largest angle. Then $\frac{1}{5}x$ represents the measure of the smallest angle, and $2\left(\frac{1}{5}x\right) = \frac{2}{5}x$ represents the measure of the other angle. Since the sum of the three angles is 180°, the equation is:
$$x + \frac{1}{5}x + \frac{2}{5}x = 180°$$
$$\frac{5}{5}x + \frac{1}{5}x + \frac{2}{5}x = 180°$$
$$\frac{8}{5}x = 180°$$
$$x = 112.5°$$
$$\frac{1}{5}x = 22.5°$$
$$\frac{2}{5}x = 45°$$
The measures of the three angles are 22.5°, 45°, and 112.5°.

11.   Let x represent the measure of the other acute angle, and 90° is the measure of the right angle. Since the sum of the three angles is 180°, the equation is:
$$x + 37° + 90° = 180°$$
$$x + 127° = 180°$$
$$x = 53°$$
The other two angles are 53° and 90°.

13.   Let x represent the total minutes for the call. Then $0.41 is charged for the first minute, and $0.32 is charged for the additional $x - 1$ minutes. The equation is:
$$0.41(1) + 0.32(x - 1) = 5.21$$
$$0.41 + 0.32x - 0.32 = 5.21$$
$$0.32x + 0.09 = 5.21$$
$$0.32x = 5.12$$
$$x = 16$$
The call was 16 minutes long.

15.   Let x represent the hours Jo Ann worked that week. Then $12/hour is paid for the first 35 hours and $18/hour is paid for the additional $x - 35$ hours. The equation is:
$$12(35) + 18(x - 35) = 492$$
$$420 + 18x - 630 = 492$$
$$18x - 210 = 492$$
$$18x = 702$$
$$x = 39$$
Jo Ann worked 39 hours that week.

17. Let x represent the number of children's tickets Stacey sold, so $2x$ represents the number of adult tickets sold. The equation is:
$$6.00(2x) + 4.50(x) = 115.50$$
$$12x + 4.5x = 115.5$$
$$16.5x = 115.5$$
$$x = 7$$
$$2x = 14$$
Stacey sold 7 children's tickets and 14 adult tickets.

19. For Jeff, the total time traveled is $\dfrac{425 \text{ miles}}{55 \text{ miles / hour}} \approx 7.72 \text{ hours} \approx 463 \text{ minutes}$. Since he left at 11:00 AM, he will arrive at 6:43 PM. For Carla, the total time traveled is $\dfrac{425 \text{ miles}}{65 \text{ miles / hour}} \approx 6.54 \text{ hours} \approx 392 \text{ minutes}$. Since she left at 1:00 PM, she will arrive at 7:32 PM. Thus Jeff will arrive in Lake Tahoe first.

21. Since $\frac{1}{5}$ mile $= 0.2$ mile, the taxi charge is \$1.25 for the first $\frac{1}{5}$ mile and \$0.25 per fifth mile for the remaining 7.3 miles. Since $7.3 \text{ miles} = \dfrac{7.3}{0.2} = 36.5 \text{ fifths}$, the total charge is: $\$1.25 + \$0.25(36.5) \approx \$10.38$

23. The first $\frac{1}{5}$ mile is \$1.25, and the remaining $12.4 - 0.2 = 12.2$ miles will be charged at \$0.25 per fifth mile. Since $12.2 \text{ miles} = \dfrac{12.2}{0.2} = 61 \text{ fifths}$, the total charge is: $\$1.25 + \$0.25(61) = \$16.50$. Yes, the meter is working correctly.

25. If all 36 people are Elk's Lodge members (which would be the least amount), the cost of the lessons would be $\$3(36) = \$108$. Since half of the money is paid to Ike and Nancy, the least amount they could make is $\frac{1}{2}(\$108) = \$54$.

27. Yes. The total receipts were \$160, which is possible if there were 10 Elk's members and 26 nonmembers. Computing the total receipts: $10(\$3) + 26(\$5) = \$30 + \$130 = \$160$

29. Let x represent the median starting salary for 1998 graduates. The equation is:
$$x + 0.07x = 85,200$$
$$1.07x = 85,200$$
$$x = \frac{85,200}{1.07} \approx \$79,626$$
The median starting salary for 1998 graduates was approximately \$79,626.

31. The pattern is to add -4, so the next number is: $-4 + (-4) = -8$

33. The pattern is to multiply by $-\frac{1}{2}$, so the next number is: $-\frac{3}{2}\left(-\frac{1}{2}\right) = \frac{3}{4}$

35. Each number is the square of a number, with alternating signs. For example, $1^2 = 1$, $-2^2 = -4$, $3^2 = 9$, and $-4^2 = -16$. Based on this pattern, the next number is: $5^2 = 25$

37. The pattern is to add $-\frac{1}{2}$, so the next number is: $\frac{1}{2} + \left(-\frac{1}{2}\right) = 0$

2.8 Linear Inequalities

1. Solving the inequality:
$$x - 5 < 7$$
$$x - 5 + 5 < 7 + 5$$
$$x < 12$$
Graphing the solution set:

3. Solving the inequality:
$$a - 4 \le 8$$
$$a - 4 + 4 \le 8 + 4$$
$$a \le 12$$
Graphing the solution set:

5. Solving the inequality:
$$x - 4.3 > 8.7$$
$$x - 4.3 + 4.3 > 8.7 + 4.3$$
$$x > 13$$
Graphing the solution set:

7. Solving the inequality:
$$y + 6 \ge 10$$
$$y + 6 + (-6) \ge 10 + (-6)$$
$$y \ge 4$$
Graphing the solution set:

9. Solving the inequality:
$$2 < x - 7$$
$$2 + 7 < x - 7 + 7$$
$$9 < x$$
$$x > 9$$
Graphing the solution set:

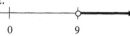

11. Solving the inequality:
$$3x < 6$$
$$\tfrac{1}{3}(3x) < \tfrac{1}{3}(6)$$
$$x < 2$$
Graphing the solution set:

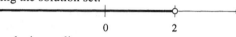

13. Solving the inequality:
$$5a \le 25$$
$$\tfrac{1}{5}(5a) \le \tfrac{1}{5}(25)$$
$$a \le 5$$
Graphing the solution set:

15. Solving the inequality:
$$\frac{x}{3} > 5$$
$$3\left(\frac{x}{3}\right) > 3(5)$$
$$x > 15$$
Graphing the solution set:

17. Solving the inequality:
$$-2x > 6$$
$$-\tfrac{1}{2}(-2x) < -\tfrac{1}{2}(6)$$
$$x < -3$$
Graphing the solution set:

19. Solving the inequality:
$$-3x \ge -18$$
$$-\tfrac{1}{3}(-3x) \le -\tfrac{1}{3}(-18)$$
$$x \le 6$$
Graphing the solution set:

21. Solving the inequality:
$$-\frac{x}{5} \le 10$$
$$-5\left(-\frac{x}{5}\right) \ge -5(10)$$
$$x \ge -50$$
Graphing the solution set:

23. Solving the inequality:
$$-\tfrac{2}{3}y > 4$$
$$-\tfrac{3}{2}\left(-\tfrac{2}{3}y\right) < -\tfrac{3}{2}(4)$$
$$y < -6$$
Graphing the solution set:

25. Solving the inequality:
$$2x - 3 < 9$$
$$2x - 3 + 3 < 9 + 3$$
$$2x < 12$$
$$\tfrac{1}{2}(2x) < \tfrac{1}{2}(12)$$
$$x < 6$$
Graphing the solution set:

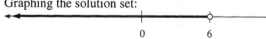

27. Solving the inequality:
$$-\tfrac{1}{5}y - \tfrac{1}{3} \le \tfrac{2}{3}$$
$$-\tfrac{1}{5}y - \tfrac{1}{3} + \tfrac{1}{3} \le \tfrac{2}{3} + \tfrac{1}{3}$$
$$-\tfrac{1}{5}y \le 1$$
$$-5\left(-\tfrac{1}{5}y\right) \ge -5(1)$$
$$y \ge -5$$
Graphing the solution set:

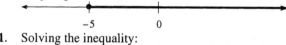

29. Solving the inequality:
$$-4x + 1 > -11$$
$$-4x + 1 + (-1) > -11 + (-1)$$
$$-4x > -12$$
$$-\tfrac{1}{4}(-4x) < -\tfrac{1}{4}(-12)$$
$$x < 3$$
Graphing the solution set:

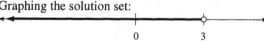

31. Solving the inequality:
$$\tfrac{2}{3}x - 5 \le 7$$
$$\tfrac{2}{3}x - 5 + 5 \le 7 + 5$$
$$\tfrac{2}{3}x \le 12$$
$$\tfrac{3}{2}\left(\tfrac{2}{3}x\right) \le \tfrac{3}{2}(12)$$
$$x \le 18$$
Graphing the solution set:

33. Solving the inequality:
$$-\tfrac{2}{5}a - 3 > 5$$
$$\tfrac{2}{5}a - 3 + 3 > 5 + 3$$
$$-\tfrac{2}{5}a > 8$$
$$-\tfrac{5}{2}\left(-\tfrac{2}{5}a\right) < -\tfrac{5}{2}(8)$$
$$a < -20$$
Graphing the solution set:

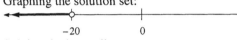

35. Solving the inequality:
$$5 - \tfrac{3}{5}y > -10$$
$$-5 + 5 - \tfrac{3}{5}y > -5 + (-10)$$
$$-\tfrac{3}{5}y > -15$$
$$-\tfrac{5}{3}\left(-\tfrac{3}{5}y\right) < -\tfrac{5}{3}(-15)$$
$$y < 25$$
Graphing the solution set:

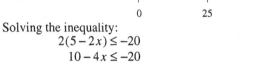

37. Solving the inequality:
$$0.3(a + 1) \le 1.2$$
$$0.3a + 0.3 \le 1.2$$
$$0.3a + 0.3 + (-0.3) \le 1.2 + (-0.3)$$
$$0.3a \le 0.9$$
$$\frac{0.3a}{0.3} \le \frac{0.9}{0.3}$$
$$a \le 3$$
Graphing the solution set:

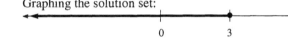

39. Solving the inequality:
$$2(5 - 2x) \le -20$$
$$10 - 4x \le -20$$
$$-10 + 10 - 4x \le -10 + (-20)$$
$$-4x \le -30$$
$$-\tfrac{1}{4}(-4x) \ge -\tfrac{1}{4}(-30)$$
$$x \ge \tfrac{15}{2}$$
Graphing the solution set:

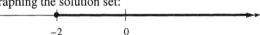

41. Solving the inequality:

$$3x - 5 > 8x$$
$$-3x + 3x - 5 > -3x + 8x$$
$$-5 > 5x$$
$$\tfrac{1}{5}(-5) > \tfrac{1}{5}(5x)$$
$$-1 > x$$
$$x < -1$$

Graphing the solution set:

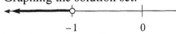

43. Solving the inequality:
$$\tfrac{1}{3}y - \tfrac{1}{2} \le \tfrac{5}{6}y + \tfrac{1}{2}$$
$$6\left(\tfrac{1}{3}y - \tfrac{1}{2}\right) \le 6\left(\tfrac{5}{6}y + \tfrac{1}{2}\right)$$
$$2y - 3 \le 5y + 3$$
$$-5y + 2y - 3 \le -5y + 5y + 3$$
$$-3y - 3 \le 3$$
$$-3y - 3 + 3 \le 3 + 3$$
$$-3y \le 6$$
$$-\tfrac{1}{3}(-3y) \ge -\tfrac{1}{3}(6)$$
$$y \ge -2$$
Graphing the solution set:

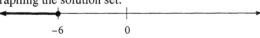

45. Solving the inequality:

$$-0.4x + 1.2 < -2x - 0.4$$
$$2x - 0.4x + 1.2 < 2x - 2x - 0.4$$
$$1.6x + 1.2 < -0.4$$
$$1.6x + 1.2 + (-1.2) < -0.4 + (-1.2)$$
$$1.6x < -1.6$$
$$\frac{1.6x}{1.6} < \frac{-1.6}{1.6}$$
$$x < -1$$

Graphing the solution set:

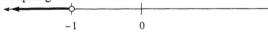

47. Solving the inequality:
$$3(m - 2) - 4 \ge 7m + 14$$
$$3m - 6 - 4 \ge 7m + 14$$
$$3m - 10 \ge 7m + 14$$
$$-7m + 3m - 10 \ge -7m + 7m + 14$$
$$-4m - 10 \ge 14$$
$$-4m - 10 + 10 \ge 14 + 10$$
$$-4m \ge 24$$
$$-\tfrac{1}{4}(-4m) \le -\tfrac{1}{4}(24)$$
$$m \le -6$$
Graphing the solution set:

49. Solving the inequality:
$$3 - 4(x - 2) \leq -5x + 6$$
$$3 - 4x + 8 \leq -5x + 6$$
$$-4x + 11 \leq -5x + 6$$
$$-4x + 5x + 11 \leq -5x + 5x + 6$$
$$x + 11 \leq 6$$
$$x + 11 + (-11) \leq 6 + (-11)$$
$$x \leq -5$$
Graphing the solution set:

51. Solving for y:
$$3x + 2y < 6$$
$$2y < -3x + 6$$
$$y < -\frac{3}{2}x + 3$$

53. Solving for y:
$$2x - 5y > 10$$
$$-5y > -2x + 10$$
$$y < \frac{2}{5}x - 2$$

55. Solving for y:
$$-3x + 7y \leq 21$$
$$7y \leq 3x + 21$$
$$y \leq \frac{3}{7}x + 3$$

57. Solving for y:
$$2x - 4y \geq -4$$
$$-4y \geq -2x - 4$$
$$y \leq \frac{1}{2}x + 1$$

59. The inequality is $x < 3$.

61. The inequality is $x \geq 3$.

63. Let x and $x + 1$ represent the integers. Solving the inequality:
$$x + x + 1 \geq 583$$
$$2x + 1 \geq 583$$
$$2x \geq 582$$
$$x \geq 291$$
The two numbers are at least 291.

65. Let x represent the number. Solving the inequality:
$$2x + 6 < 10$$
$$2x < 4$$
$$x < 2$$

67. Let x represent the number. Solving the inequality:
$$4x > x - 8$$
$$3x > -8$$
$$x > -\frac{8}{3}$$

69. Let w represent the width, so $3w$ represents the length. Using the formula for perimeter:
$$2(w) + 2(3w) \geq 48$$
$$2w + 6w \geq 48$$
$$8w \geq 48$$
$$w \geq 6$$
The width is at least 6 meters.

71. Let x, $x + 2$, and $x + 4$ represent the sides of the triangle. The inequality is:
$$x + (x + 2) + (x + 4) > 24$$
$$3x + 6 > 24$$
$$3x > 18$$
$$x > 6$$
The shortest side is an even number greater than 6 inches (greater than or equal to 8 inches).

73. The inequality is $t \geq 100$.

75. Let n represent the number of tickets they sell. The inequality is:
$$7.5n < 1500$$
$$n < 200$$
They will lose money if they sell less than 200 tickets. They will make a profit if they sell more than 200 tickets.

77. b (commutative property of addition)

79. a (distributive property)

81. b and c (commutative and associative properties of addition)

2.9 Compound Inequalities

1. Graphing the solution set:

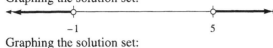

3. Graphing the solution set:

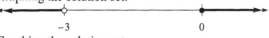

5. Graphing the solution set:

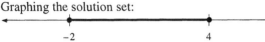

7. Graphing the solution set:

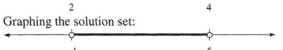

9. Graphing the solution set:

11. Graphing the solution set:

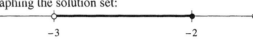

13. Graphing the solution set:

15. Graphing the solution set:

17. Solving the compound inequality:

$$3x - 1 < 5 \qquad \text{or} \qquad 5x - 5 > 10$$
$$3x < 6 \qquad\qquad\qquad 5x > 15$$
$$x < 2 \qquad\qquad\qquad x > 3$$

Graphing the solution set:

19. Solving the compound inequality:

$$x - 2 > -5 \qquad \text{and} \qquad x + 7 < 13$$
$$x > -3 \qquad\qquad\qquad x < 6$$

Graphing the solution set:

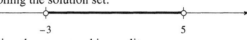

21. Solving the compound inequality:

$$11x < 22 \qquad \text{or} \qquad 12x > 36$$
$$x < 2 \qquad\qquad\qquad x > 3$$

Graphing the solution set:

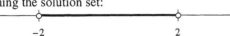

23. Solving the compound inequality:

$$3x - 5 < 10 \qquad \text{and} \qquad 2x + 1 > -5$$
$$3x < 15 \qquad\qquad\qquad 2x > -6$$
$$x < 5 \qquad\qquad\qquad x > -3$$

Graphing the solution set:

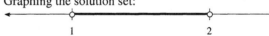

25. Solving the compound inequality:

$$2x - 3 < 8 \qquad \text{and} \qquad 3x + 1 > -10$$
$$2x < 11 \qquad\qquad\qquad 3x > -11$$
$$x < \frac{11}{2} \qquad\qquad\qquad x > -\frac{11}{3}$$

Graphing the solution set:

27. Solving the compound inequality:

$$2x - 1 < 3 \qquad \text{and} \qquad 3x - 2 > 1$$
$$2x < 4 \qquad\qquad\qquad 3x > 3$$
$$x < 2 \qquad\qquad\qquad x > 1$$

Graphing the solution set:

29. Solving the compound inequality:

$$-1 \le x - 5 \le 2$$
$$4 \le x \le 7$$

Graphing the solution set:

31. Solving the compound inequality:

$$-4 \le 2x \le 6$$
$$-2 \le x \le 3$$

Graphing the solution set:

33. Solving the compound inequality:

$$-3 < 2x + 1 < 5$$
$$-4 < 2x < 4$$
$$-2 < x < 2$$

Graphing the solution set:

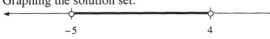

35. Solving the compound inequality:

$$0 \le 3x + 2 \le 7$$
$$-2 \le 3x \le 5$$
$$-\frac{2}{3} \le x \le \frac{5}{3}$$

Graphing the solution set:

37. Solving the compound inequality:

$$-7 < 2x + 3 < 11$$
$$-10 < 2x < 8$$
$$-5 < x < 4$$

Graphing the solution set:

39. Solving the compound inequality:

$$-1 \le 4x + 5 \le 9$$
$$-6 \le 4x \le 4$$
$$-\frac{3}{2} \le x \le 1$$

Graphing the solution set:

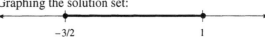

41.   The inequality is $-2 < x < 3$.   **43.**   The inequality is $-2 \le x \le 3$.

45.   **a.**   The three inequalities are $2x + x > 10$, $2x + 10 > x$, and $x + 10 > 2x$.

         b.   For $x + 10 > 2x$ we have $x < 10$. For $3x > 10$ we have $x > \frac{10}{3}$. The compound inequality is $\frac{10}{3} < x < 10$.

47.   Graphing the inequality:

50   266

49.   Let x represent the number. Solving the inequality:

$$5 < 2x - 3 < 7$$
$$8 < 2x < 10$$
$$4 < x < 5$$

The number is between 4 and 5.

51.   Let w represent the width and $w + 4$ represent the length. Using the perimeter formula:

$$20 < 2w + 2(w + 4) < 30$$
$$20 < 2w + 2w + 8 < 30$$
$$20 < 4w + 8 < 30$$
$$12 < 4w < 22$$
$$3 < w < \frac{11}{2}$$

The width is between 3 inches and $\frac{11}{2} = 5\frac{1}{2}$ inches.

53.   Simplifying the expression: $-|-5| = -(5) = -5$

55.   Simplifying the expression: $-3 - 4(-2) = -3 + 8 = 5$

57.   Simplifying the expression: $5|3 - 8| - 6|2 - 5| = 5|-5| - 6|-3| = 5(5) - 6(3) = 25 - 18 = 7$

59.   Simplifying the expression: $5 - 2[-3(5 - 7) - 8] = 5 - 2[-3(-2) - 8] = 5 - 2(6 - 8) = 5 - 2(-2) = 5 + 4 = 9$

61.   The expression is: $-3 - (-9) = -3 + 9 = 6$

63.   Applying the distributive property: $\frac{1}{2}(4x - 6) = \frac{1}{2} \cdot 4x - \frac{1}{2} \cdot 6 = 2x - 3$

65.   The integers are: $-3, 0, 2$

Chapter 2 Review

1.   Simplifying the expression: $5x - 8x = (5 - 8)x = -3x$

3.   Simplifying the expression: $-a + 2 + 5a - 9 = -a + 5a + 2 - 9 = 4a - 7$

5.   Simplifying the expression: $6 - 2(3y + 1) - 4 = 6 - 6y - 2 - 4 = -6y + 6 - 2 - 4 = -6y$

7.   Evaluating when $x = 3$: $7x - 2 = 7(3) - 2 = 21 - 2 = 19$

9.   Evaluating when $x = 3$: $-x - 2x - 3x = -6x = -6(3) = -18$

11.   Evaluating when $x = -2$: $-3x + 2 = -3(-2) + 2 = 6 + 2 = 8$

13.   Solving the equation: **15.**   Solving the equation:

$$x + 2 = -6$$
$$x + 2 + (-2) = -6 + (-2)$$
$$x = -8$$

$$10 - 3y + 4y = 12$$
$$10 + y = 12$$
$$-10 + 10 + y = -10 + 12$$
$$y = 2$$

17.   Solving the equation: **19.**   Solving the equation:

$$2x = -10$$
$$\frac{1}{2}(2x) = \frac{1}{2}(-10)$$
$$x = -5$$

$$\frac{x}{3} = 4$$
$$3\left(\frac{x}{3}\right) = 3(4)$$
$$x = 12$$

21.   Solving the equation:

$$3a - 2 = 5a$$
$$-3a + 3a - 2 = -3a + 5a$$
$$-2 = 2a$$
$$\tfrac{1}{2}(-2) = \tfrac{1}{2}(2a)$$
$$a = -1$$

23.   Solving the equation:

$$3x + 2 = 5x - 8$$
$$3x + (-5x) + 2 = 5x + (-5x) - 8$$
$$-2x + 2 = -8$$
$$-2x + 2 + (-2) = -8 + (-2)$$
$$-2x = -10$$
$$-\tfrac{1}{2}(-2x) = -\tfrac{1}{2}(-10)$$
$$x = 5$$

25.   Solving the equation:

$$0.7x - 0.1 = 0.5x - 0.1$$
$$0.7x + (-0.5x) - 0.1 = 0.5x + (-0.5x) - 0.1$$
$$0.2x - 0.1 = -0.1$$
$$0.2x - 0.1 + 0.1 = -0.1 + 0.1$$
$$0.2x = 0$$
$$\frac{0.2x}{0.2} = \frac{0}{0.2}$$
$$x = 0$$

27.   Simplifying and then solving the equation:

$$2(x - 5) = 10$$
$$2x - 10 = 10$$
$$2x - 10 + 10 = 10 + 10$$
$$2x = 20$$
$$\tfrac{1}{2}(2x) = \tfrac{1}{2}(20)$$
$$x = 10$$

29.   Simplifying and then solving the equation:

$$\tfrac{1}{2}(3t - 2) + \tfrac{1}{2} = \tfrac{5}{2}$$
$$\tfrac{3}{2}t - 1 + \tfrac{1}{2} = \tfrac{5}{2}$$
$$\tfrac{3}{2}t - \tfrac{1}{2} = \tfrac{5}{2}$$
$$\tfrac{3}{2}t - \tfrac{1}{2} + \tfrac{1}{2} = \tfrac{5}{2} + \tfrac{1}{2}$$
$$\tfrac{3}{2}t = 3$$
$$\tfrac{2}{3}\left(\tfrac{3}{2}t\right) = \tfrac{2}{3}(3)$$
$$t = 2$$

31.   Simplifying and then solving the equation:

$$2(3x + 7) = 4(5x - 1) + 18$$
$$6x + 14 = 20x - 4 + 18$$
$$6x + 14 = 20x + 14$$
$$6x + (-20x) + 14 = 20x + (-20x) + 14$$
$$-14x + 14 = 14$$
$$-14x + 14 + (-14) = 14 + (-14)$$
$$-14x = 0$$
$$-\tfrac{1}{14}(-14x) = -\tfrac{1}{14}(0)$$
$$x = 0$$

33.   Substituting $x = 5$:

$$4(5) - 5y = 20$$
$$20 - 5y = 20$$
$$-5y = 0$$
$$y = 0$$

35.   Substituting $x = -5$:

$$4(-5) - 5y = 20$$
$$-20 - 5y = 20$$
$$-5y = 40$$
$$y = -8$$

37.   Solving for y:

$$2x - 5y = 10$$
$$-5y = -2x + 10$$
$$y = \tfrac{2}{5}x - 2$$

39.   Solving for h:

$$\pi r^2 h = V$$
$$\frac{\pi r^2 h}{\pi r^2} = \frac{V}{\pi r^2}$$
$$h = \frac{V}{\pi r^2}$$

41.   Computing the amount:

$$0.86(240) = x$$
$$x = 206.4$$

So 86% of 240 is 206.4.

43.   Let x represent the number. The equation is:

$$2x + 6 = 28$$
$$2x = 22$$
$$x = 11$$

The number is 11.

45. Completing the table:

	Dollars Invested at 9%	Dollars Invested at 10%
Number of	x	$x + 300$
Interest on	$0.09(x)$	$0.10(x + 300)$

The equation is:
$$0.09(x) + 0.10(x + 300) = 125$$
$$0.09x + 0.10x + 30 = 125$$
$$0.19x + 30 = 125$$
$$0.19x = 95$$
$$x = 500$$
$$x + 300 = 800$$
The man invested $500 at 9% and $800 at 10%.

47. Solving the inequality:
$$-2x < 4$$
$$-\tfrac{1}{2}(-2x) > -\tfrac{1}{2}(4)$$
$$x > -2$$

49. Solving the inequality:
$$-\frac{a}{2} \le -3$$
$$-2\left(-\frac{a}{2}\right) \ge -2(-3)$$
$$a \ge 6$$

51. Solving the inequality:
$$-4x + 5 > 37$$
$$-4x > 32$$
$$x < -8$$

Graphing the solution set:

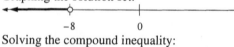

$$-8 \qquad 0$$

53. Solving the inequality:
$$2(3t + 1) + 6 \ge 5(2t + 4)$$
$$6t + 2 + 6 \ge 10t + 20$$
$$6t + 8 \ge 10t + 20$$
$$-4t + 8 \ge 20$$
$$-4t \ge 12$$
$$t \le -3$$

Graphing the solution set:

$$-3 \qquad 0$$

55. Solving the compound inequality:

$$-5x \ge 25 \qquad \text{or} \qquad 2x - 3 \ge 9$$
$$x \le -5 \qquad\qquad\qquad 2x \ge 12$$
$$x \le -5 \qquad\qquad\qquad x \ge 6$$

Graphing the solution set:

$$-5 \qquad\qquad\qquad 6$$

Chapters 1-2 Cumulative Review

1. Simplifying: $6 + 3(6 + 2) = 6 + 3(8) = 6 + 24 = 30$

3. Simplifying: $7 - 9 - 12 = 7 + (-9) + (-12) = 7 + (-21) = -14$

5. Using the associative property: $\frac{1}{5}(10x) = \left(\frac{1}{5} \cdot 10\right)x = 2x$

7. Simplifying: $\left(-\frac{2}{3}\right)^3 = \left(-\frac{2}{3}\right)\left(-\frac{2}{3}\right)\left(-\frac{2}{3}\right) = -\frac{8}{27}$

9. Simplifying: $-\frac{3}{4} \div \frac{15}{16} = -\frac{3}{4} \cdot \frac{16}{15} = -\frac{49}{60} = -\frac{4}{5}$

11. Simplifying: $\dfrac{-4(-6)}{-9} = \dfrac{24}{-9} = -\frac{8}{3}$

13. Simplifying: $\dfrac{(5-3)^2}{5^2 - 3^2} = \dfrac{2^2}{25 - 9} = \frac{4}{16} = \frac{1}{4}$

15. First factor the denominators to find the LCM:
$$21 = 3 \cdot 7$$
$$35 = 5 \cdot 7$$
$$\text{LCM} = 3 \cdot 5 \cdot 7 = 105$$
Simplifying: $\dfrac{4}{21} - \dfrac{9}{35} = \dfrac{4 \cdot 5}{21 \cdot 5} - \dfrac{9 \cdot 3}{35 \cdot 3} = \dfrac{20}{105} - \dfrac{27}{105} = -\dfrac{7}{105} = -\dfrac{7}{3 \cdot 5 \cdot 7} = -\dfrac{1}{3 \cdot 5} = -\dfrac{1}{15}$

17. Solving the equation:

$$7x = 6x + 4$$
$$7x + (-6x) = 6x + (-6x) + 4$$
$$x = 4$$

19. Solving the equation:

$$-\tfrac{3}{5}x = 30$$
$$-\tfrac{5}{3}\left(-\tfrac{3}{5}x\right) = -\tfrac{5}{3}(30)$$
$$x = -50$$

21. Solving the equation:

$$5x - 7 = x - 1$$
$$5x + (-x) - 7 = x + (-x) - 1$$
$$4x - 7 = -1$$
$$4x - 7 + 7 = -1 + 7$$
$$4x = 6$$
$$\tfrac{1}{4}(4x) = \tfrac{1}{4}(6)$$
$$x = \tfrac{3}{2}$$

23. Solving the equation:

$$15 - 3(2t + 4) = 1$$
$$15 - 6t - 12 = 1$$
$$-6t + 3 = 1$$
$$-6t + 3 + (-3) = 1 + (-3)$$
$$-6t = -2$$
$$-\tfrac{1}{6}(-6t) = -\tfrac{1}{6}(-2)$$
$$t = \tfrac{1}{3}$$

25. Solving the equation:

$$\tfrac{1}{3}(x - 6) = \tfrac{1}{4}(x + 8)$$
$$\tfrac{1}{3}x - 2 = \tfrac{1}{4}x + 2$$
$$12\left(\tfrac{1}{3}x - 2\right) = 12\left(\tfrac{1}{4}x + 2\right)$$
$$4x - 24 = 3x + 24$$
$$4x + (-3x) - 24 = 3x + (-3x) + 24$$
$$x - 24 = 24$$
$$x - 24 + 24 = 24 + 24$$
$$x = 48$$

27. Solving for y:

$$3x + 4y = 12$$
$$3x + (-3x) + 4y = 12 + (-3x)$$
$$4y = -3x + 12$$
$$\tfrac{1}{4}(4y) = \tfrac{1}{4}(-3x + 12)$$
$$y = -\tfrac{3}{4}x + 3$$

29. Solving the inequality:

$$-5x + 9 < -6$$
$$-5x + 9 + (-9) < -6 + (-9)$$
$$-5x < -15$$
$$-\tfrac{1}{5}(-5x) > -\tfrac{1}{5}(-15)$$
$$x > 3$$

Graphing the solution set:

31. Solving the compound inequality:

$$-2 < x + 1 < 5$$
$$-3 < x < 4$$

Graphing the solution set:

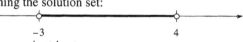

33. The opposite of $-\tfrac{2}{3}$ is $\tfrac{2}{3}$, the reciprocal is $-\tfrac{3}{2}$, and the absolute value is $\left|-\tfrac{2}{3}\right| = \tfrac{2}{3}$.

35. The pattern is to add -3, so the next term is: $-5 + (-3) = -8$

37. Using the distributive property: $\tfrac{1}{4}(8x - 4) = \tfrac{1}{4} \cdot 8x - \tfrac{1}{4} \cdot 4 = 2x - 1$

39. Reducing the fraction: $\dfrac{234}{312} = \dfrac{2 \cdot 3 \cdot 3 \cdot 13}{2 \cdot 2 \cdot 2 \cdot 3 \cdot 13} = \dfrac{3}{2 \cdot 2} = \dfrac{3}{4}$

41. Evaluating when $a = 3$ and $b = -2$: $a^2 - 2ab + b^2 = (3)^2 - 2(3)(-2) + (-2)^2 = 9 + 12 + 4 = 25$

43. Let x represent the number. The equation is:

$$2x + 7 = 31$$
$$2x = 24$$
$$x = 12$$

The number is 12.

45. Let x represent the acute angle and $90°$ is the right angle. Since the sum of the three angles is $180°$, the equation is:

$$x + 42° + 90° = 180°$$
$$x + 132° = 180°$$
$$x = 48°$$

The other two angles are $48°$ and $90°$.

47. The other angle must be $90° - 25° = 65°$.

49. Completing the table:

	Dollars Invested at 5%	Dollars Invested at 6%
Number of	x	$x+200$
Interest on	$0.05(x)$	$0.06(x+200)$

The equation is:
$$0.05(x)+0.06(x+200)=56$$
$$0.05x+0.06x+12=56$$
$$0.11x+12=56$$
$$0.11x=44$$
$$x=400$$
$$x+200=600$$
You have \$400 invested at 5% and \$600 invested at 6%.

Chapter 2 Test

1. Simplifying: $3x+2-7x+3=3x-7x+2+3=-4x+5$
2. Simplifying: $4a-5-a+1=4a-a-5+1=3a-4$
3. Simplifying: $7-3(y+5)-4=7-3y-15-4=-3y+7-15-4=-3y-12$
4. Simplifying: $8(2x+1)-5(x-4)=16x+8-5x+20=16x-5x+8+20=11x+28$
5. Evaluating when $x=-5$: $2x-3-7x=-5x-3=-5(-5)-3=25-3=22$
6. Evaluating when $x=2$ and $y=3$: $x^2+2xy+y^2=(2)^2+2(2)(3)+(3)^2=4+12+9=25$

7. Solving the equation:
$$2x-5=7$$
$$2x-5+5=7+5$$
$$2x=12$$
$$\tfrac{1}{2}(2x)=\tfrac{1}{2}(12)$$
$$x=6$$

8. Solving the equation:
$$2y+4=5y$$
$$-2y+2y+4=-2y+5y$$
$$4=3y$$
$$\tfrac{1}{3}(4)=\tfrac{1}{3}(3y)$$
$$y=\tfrac{4}{3}$$

9. First clear the equation of fractions by multiplying by 10:
$$\tfrac{1}{2}x-\tfrac{1}{10}=\tfrac{1}{5}x+\tfrac{1}{2}$$
$$10\left(\tfrac{1}{2}x-\tfrac{1}{10}\right)=10\left(\tfrac{1}{5}x+\tfrac{1}{2}\right)$$
$$5x-1=2x+5$$
$$5x+(-2x)-1=2x+(-2x)+5$$
$$3x-1=5$$
$$3x-1+1=5+1$$
$$3x=6$$
$$\tfrac{1}{3}(3x)=\tfrac{1}{3}(6)$$
$$x=2$$

10. Solving the equation:
$$\tfrac{2}{5}(5x-10)=-5$$
$$2x-4=-5$$
$$2x-4+4=-5+4$$
$$2x=-1$$
$$\tfrac{1}{2}(2x)=\tfrac{1}{2}(-1)$$
$$x=-\tfrac{1}{2}$$

11. Solving the equation:
$$-5(2x+1)-6=19$$
$$-10x-5-6=19$$
$$-10x-11=19$$
$$-10x-11+11=19+11$$
$$-10x=30$$
$$-\tfrac{1}{10}(-10x)=-\tfrac{1}{10}(30)$$
$$x=-3$$

12. Solving the equation:

$$0.04x + 0.06(100 - x) = 4.6$$
$$0.04x + 6 - 0.06x = 4.6$$
$$-0.02x + 6 = 4.6$$
$$-0.02x + 6 + (-6) = 4.6 + (-6)$$
$$-0.02x = -1.4$$
$$\frac{-0.02x}{-0.02} = \frac{-1.4}{-0.02}$$
$$x = 70$$

13. Solving the equation:

$$2(t - 4) + 3(t + 5) = 2t - 2$$
$$2t - 8 + 3t + 15 = 2t - 2$$
$$5t + 7 = 2t - 2$$
$$5t + (-2t) + 7 = 2t + (-2t) - 2$$
$$3t + 7 = -2$$
$$3t + 7 + (-7) = -2 + (-7)$$
$$3t = -9$$
$$\tfrac{1}{3}(3t) = \tfrac{1}{3}(-9)$$
$$t = -3$$

14. Solving the equation:

$$2x - 4(5x + 1) = 3x + 17$$
$$2x - 20x - 4 = 3x + 17$$
$$-18x - 4 = 3x + 17$$
$$-18x + (-3x) - 4 = 3x + (-3x) + 17$$
$$-21x - 4 = 17$$
$$-21x - 4 + 4 = 17 + 4$$
$$-21x = 21$$
$$-\tfrac{1}{21}(21x) = -\tfrac{1}{21}(21)$$
$$x = -1$$

15. Finding the amount:

$$0.15(38) = x$$
$$5.7 = x$$

So 15% of 38 is 5.7.

16. Finding the base:

$$0.12x = 240$$
$$\frac{0.12x}{0.12} = \frac{240}{0.12}$$
$$x = 2000$$

So 12% of 2,000 is 240.

17. Substituting $y = -2$:

$$2x - 3(-2) = 12$$
$$2x + 6 = 12$$
$$2x = 6$$
$$x = 3$$

18. Substituting $V = 88$, $\pi = \frac{22}{7}$, and $r = 3$:

$$\tfrac{1}{3} \cdot \tfrac{22}{7} \cdot (3)^2 h = 88$$
$$\tfrac{66}{7} h = 88$$
$$\tfrac{7}{66}\left(\tfrac{66}{7} h\right) = \tfrac{7}{66}(88)$$
$$h = \tfrac{28}{3} \text{ inches}$$

19. Solving for y:

$$2x + 5y = 20$$
$$5y = -2x + 20$$
$$y = -\tfrac{2}{5}x + 4$$

20. Solving for v:

$$x + vt + 16t^2 = h$$
$$vt = h - x - 16t^2$$
$$v = \frac{h - x - 16t^2}{t}$$

21. Completing the table:

	Ten Years Ago	Now
Dave	$2x - 10$	$2x$
Rick	$x - 10$	x

The equation is:

$$2x - 10 + x - 10 = 40$$
$$3x - 20 = 40$$
$$3x = 60$$
$$x = 20$$
$$2x = 40$$

Rick is 20 years old and Dave is 40 years old.

22.   Let w represent the width and $2w$ represent the length. Using the perimeter formula:
$$2(w) + 2(2w) = 60$$
$$2w + 4w = 60$$
$$6w = 60$$
$$w = 10$$
$$2w = 20$$
The width is 10 inches and the length is 20 inches.

23.   Completing the table:

	Dimes	Quarters
Number	$x + 7$	x
Value (cents)	$10(x+7)$	$25(x)$

The equation is:
$$10(x+7) + 25(x) = 350$$
$$10x + 70 + 25x = 350$$
$$35x + 70 = 350$$
$$35x = 280$$
$$x = 8$$
$$x + 7 = 15$$
He has 8 quarters and 15 dimes in his collection.

24.   Completing the table:

	Dollars Invested at 7%	Dollars Invested at 9%
Number of	x	$x + 600$
Interest on	$0.07(x)$	$0.09(x+600)$

The equation is:
$$0.07(x) + 0.09(x+600) = 182$$
$$0.07x + 0.09x + 54 = 182$$
$$0.16x + 54 = 182$$
$$0.16x = 128$$
$$x = 800$$
$$x + 600 = 1400$$
She has $800 invested at 7% and $1,400 invested at 9%.

25.   Solving the inequality:
$$2x + 3 < 5$$
$$2x + 3 + (-3) < 5 + (-3)$$
$$2x < 2$$
$$\tfrac{1}{2}(2x) < \tfrac{1}{2}(2)$$
$$x < 1$$
Graphing the solution set:

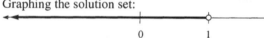

26.   Solving the inequality:
$$-5a > 20$$
$$-\tfrac{1}{5}(-5a) < -\tfrac{1}{5}(20)$$
$$a < -4$$
Graphing the solution set:

27.   Solving the inequality:
$$0.4 - 0.2x \geq 1$$
$$-0.4 + 0.4 - 0.2x \geq -0.4 + 1$$
$$-0.2x \geq 0.6$$
$$\frac{-0.2x}{-0.2} \leq \frac{0.6}{-0.2}$$
$$x \leq -3$$
Graphing the solution set:

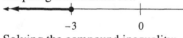

28.   Solving the inequality:
$$4 - 5(m+1) \leq 9$$
$$4 - 5m - 5 \leq 9$$
$$-5m - 1 \leq 9$$
$$-5m - 1 + 1 \leq 9 + 1$$
$$-5m \leq 10$$
$$-\tfrac{1}{5}(-5m) \geq -\tfrac{1}{5}(10)$$
$$m \geq -2$$
Graphing the solution set:

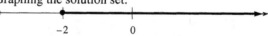

29.   Solving the compound inequality:
$$3 - 4x \geq -5 \quad \text{or} \quad 2x \geq 10$$
$$-4x \geq -8 \qquad\qquad x \geq 5$$
$$x \leq 2 \qquad\qquad\quad x \geq 5$$
Graphing the solution set:

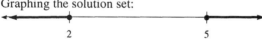

30.   Solving the compound inequality:
$$-7 < 2x - 1 < 9$$
$$-6 < 2x < 10$$
$$-3 < x < 5$$
Graphing the solution set:

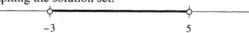

Chapter 3
Linear Equations and Inequalities in Two Variables

3.1 Paired Data and Graphing Ordered Pairs

1.   Graphing the ordered pair:

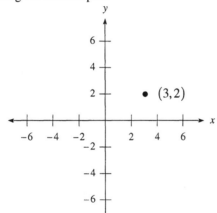

3.   Graphing the ordered pair:

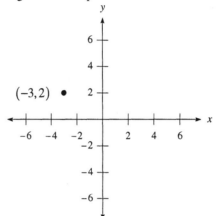

5.   Graphing the ordered pair:

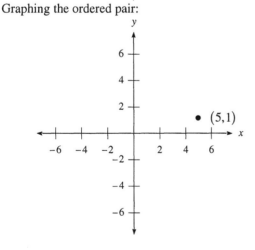

7.   Graphing the ordered pair:

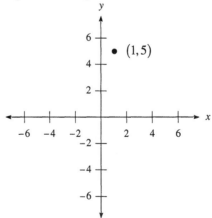

9.   Graphing the ordered pair:

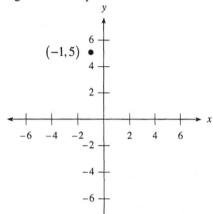

11.   Graphing the ordered pair:

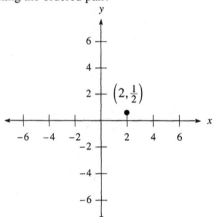

13.   Graphing the ordered pair:

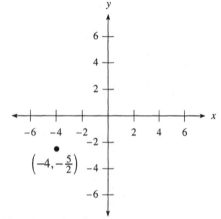

15.   Graphing the ordered pair:

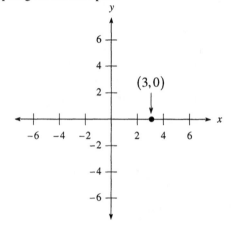

17.   Graphing the ordered pair:

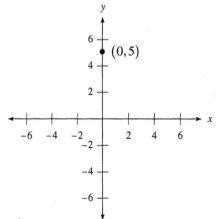

19.   The coordinates are $(-4,4)$.

21.   The coordinates are $(-4,2)$.

23.   The coordinates are $(-3,0)$.

25.   The coordinates are $(2,-2)$.

27.   The coordinates are $(-5,-5)$.

29.   Graphing the line:

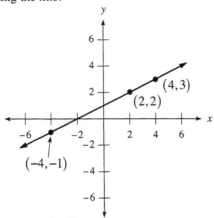

Yes, the point $(2,2)$ lies on the line.

31.   Graphing the line:

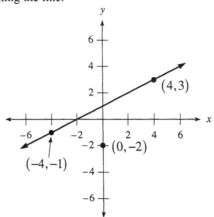

No, the point $(0,-2)$ does not lie on the line.

33.   Graphing the line:

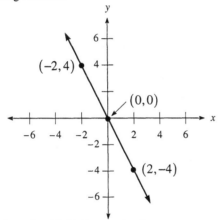

Yes, the point $(0,0)$ lies on the line.

34.   Graphing the line:

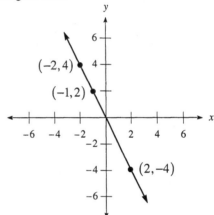

Yes, the point $(-1,2)$ lies on the line.

35.   Graphing the line:

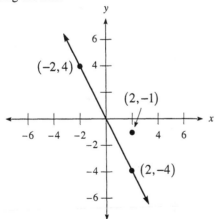

No, the point $(2,-1)$ does not lie on the line.

37.   Graphing the line:

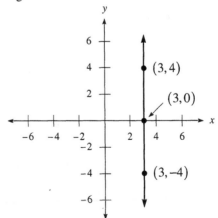

Yes, the point $(3,0)$ lies on the line.

39.   No, the x-coordinate of every point on this line is 3.

41.   Graphing the line:

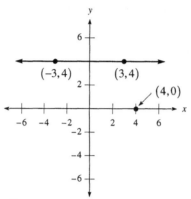

No, the point $(4,0)$ does not lie on the line.

43.   No, the y-coordinate of every point on this line is 4.

45.   **a.**   Three ordered pairs on the graph are (5,40), (10,80), and (20,160).
      b.   She will earn $320 for working 40 hours.
      c.   If her check is $240, she worked 30 hours that week.
      d.   No. She should be paid $280 for working 35 hours, not $260.

47.   Constructing a line graph:

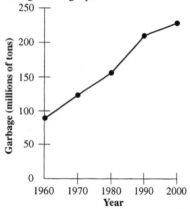

49.   Constructing a line graph:

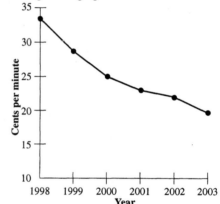

51.   Constructing a line graph:

53.   Constructing a scatter diagram:

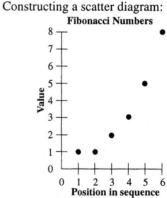

55.   The other three corners are the points $A(-2,-3)$, $B(-2,1)$, and $C(2,1)$.

57.   The pattern is to add 7, so the next number is: $24 + 7 = 31$

59.   The pattern is to multiply by $\frac{1}{3}$, so the next number is: $\frac{1}{9} \cdot \frac{1}{3} = \frac{1}{27}$

61.   The pattern is to add –3, so the next number is: $-2 + (-3) = -5$

63.   The pattern is to multiply by 3, so the next number is: $189 \cdot 3 = 567$

65.   The pattern is to add 1, add 2, add 3, so the next number is obtained by adding 4: $11 + 4 = 15$

3.2 Solutions to Linear Equations in Two Variables

1. Substituting $x = 0$, $y = 0$, and $y = -6$:

$2(0) + y = 6$	$2x + 0 = 6$	$2x + (-6) = 6$
$0 + y = 6$	$2x = 6$	$2x = 12$
$y = 6$	$x = 3$	$x = 6$

    The ordered pairs are $(0,6)$, $(3,0)$, and $(6,-6)$.

3. Substituting $x = 0$, $y = 0$, and $x = -4$:

$3(0) + 4y = 12$	$3x + 4(0) = 12$	$3(-4) + 4y = 12$
$0 + 4y = 12$	$3x + 0 = 12$	$-12 + 4y = 12$
$4y = 12$	$3x = 12$	$4y = 24$
$y = 3$	$x = 4$	$y = 6$

    The ordered pairs are $(0,3)$, $(4,0)$, and $(-4,6)$.

5. Substituting $x = 1$, $y = 0$, and $x = 5$:

$y = 4(1) - 3$	$0 = 4x - 3$	$y = 4(5) - 3$
$y = 4 - 3$	$3 = 4x$	$y = 20 - 3$
$y = 1$	$x = \frac{3}{4}$	$y = 17$

    The ordered pairs are $(1,1)$, $\left(\frac{3}{4},0\right)$, and $(5,17)$.

7. Substituting $x = 2$, $y = 6$, and $x = 0$:

$y = 7(2) - 1$	$6 = 7x - 1$	$y = 7(0) - 1$
$y = 14 - 1$	$7 = 7x$	$y = 0 - 1$
$y = 13$	$x = 1$	$y = -1$

    The ordered pairs are $(2,13)$, $(1,6)$, and $(0,-1)$.

9. Substituting $y = 4$, $y = -3$, and $y = 0$ results (in each case) in $x = -5$. The ordered pairs are $(-5,4)$, $(-5,-3)$, and $(-5,0)$.

11. Completing the table:

x	y
1	3
-3	-9
4	12
6	18

13. Completing the table:

x	y
0	0
-1/2	-2
-3	-12
3	12

15. Completing the table:

x	y
2	3
3	2
5	0
9	-4

17. Completing the table:

x	y
2	0
3	2
1	-2
-3	-10

19. Completing the table:

x	y
0	-1
-1	-7
-3	-19
3/2	8

21. Substituting each ordered pair into the equation:

    $(2,3)$: $2(2) - 5(3) = 4 - 15 = -11 \neq 10$

    $(0,-2)$: $2(0) - 5(-2) = 0 + 10 = 10$

    $\left(\frac{5}{2},1\right)$: $2\left(\frac{5}{2}\right) - 5(1) = 5 - 5 = 0 \neq 10$

    Only the ordered pair $(0,-2)$ is a solution.

23.   Substituting each ordered pair into the equation:
$(1,5)$: $7(1) - 2 = 7 - 2 = 5$
$(0,-2)$: $7(0) - 2 = 0 - 2 = -2$
$(-2,-16)$: $7(-2) - 2 = -14 - 2 = -16$
All the ordered pairs $(1,5)$, $(0,-2)$ and $(-2,-16)$ are solutions.

25.   Substituting each ordered pair into the equation:
$(1,6)$: $6(1) = 6$
$(-2,-12)$: $6(-2) = -12$
$(0,0)$: $6(0) = 0$
All the ordered pairs $(1,6)$, $(-2,-12)$ and $(0,0)$ are solutions.

27.   Substituting each ordered pair into the equation:
$(1,1)$: $1 + 1 = 2 \neq 0$
$(2,-2)$: $2 + (-2) = 0$
$(3,3)$: $3 + 3 = 6 \neq 0$
Only the ordered pair $(2,-2)$ is a solution.

29.   Since $x = 3$, the ordered pair $(5,3)$ cannot be a solution. The ordered pairs $(3,0)$ and $(3,-3)$ are solutions.

31.   Substituting $w = 3$:
$2l + 2(3) = 30$
$2l + 6 = 30$
$2l = 24$
$l = 12$
The length is 12 inches.

33.   Completing the table:

Minutes	0	10	20	30	40	50	60	70	80	90	100
Cost	$3	$4	$5	$6	$7	$8	$9	$10	$11	$12	$13

Creating a line graph:

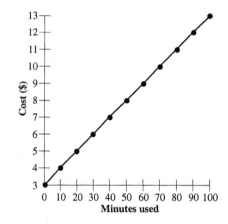

35. Completing the table:

Hours	0	1	2	3	4	5	6	7	8	9	10
Cost	$10	$13	$16	$19	$22	$25	$28	$31	$34	$37	$40

Creating a line graph:

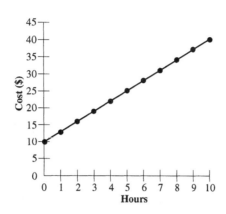

37. Substituting $x = 4$:
$$3(4) + 2y = 6$$
$$12 + 2y = 6$$
$$2y = -6$$
$$y = -3$$

39. Substituting $x = 0$: $y = -\frac{1}{3}(0) + 2 = 0 + 2 = 2$

41. Substituting $x = 2$: $y = \frac{3}{2}(2) - 3 = 3 - 3 = 0$

43. Solving for y:
$$5x + y = 4$$
$$y = -5x + 4$$

45. Solving for y:
$$3x - 2y = 6$$
$$-2y = -3x + 6$$
$$y = \frac{3}{2}x - 3$$

3.3 Graphing Linear Equations in Two Variables

1. The ordered pairs are $(0,4)$, $(2,2)$, and $(4,0)$:

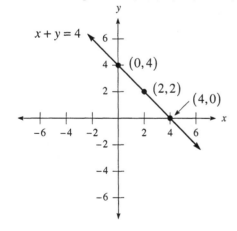

3. The ordered pairs are $(0,3)$, $(2,1)$, and $(4,-1)$:

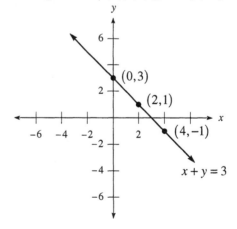

5.   The ordered pairs are $(0,0)$, $(-2,-4)$, and $(2,4)$:

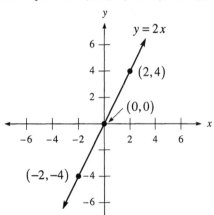

7.   The ordered pairs are $(-3,-1)$, $(0,0)$, and $(3,1)$:

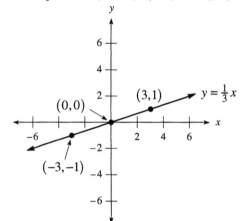

9.   The ordered pairs are $(0,1)$, $(-1,-1)$, and $(1,3)$:

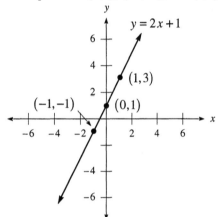

11.   The ordered pairs are $(0,4)$, $(-1,4)$, and $(2,4)$:

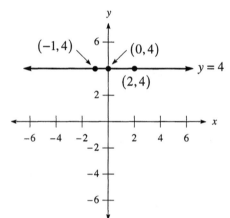

13.   The ordered pairs are $(-2,2)$, $(0,3)$, and $(2,4)$:

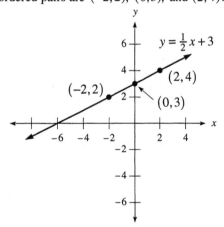

15.   The ordered pairs are $(-3,3)$, $(0,1)$, and $(3,-1)$:

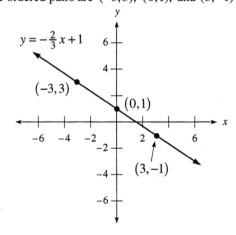

17. Solving for y:

$$2x + y = 3$$
$$y = -2x + 3$$

The ordered pairs are $(-1, 5)$, $(0, 3)$, and $(1, 1)$:

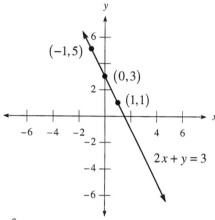

19. Solving for y:

$$3x + 2y = 6$$
$$2y = -3x + 6$$
$$y = -\frac{3}{2}x + 3$$

The ordered pairs are $(0, 3)$, $(2, 0)$, and $(4, -3)$:

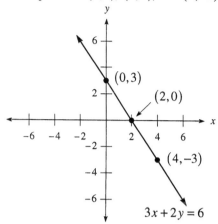

21. Solving for y:

$$-x + 2y = 6$$
$$2y = x + 6$$
$$y = \frac{1}{2}x + 3$$

The ordered pairs are $(-2, 2)$, $(0, 3)$, and $(2, 4)$:

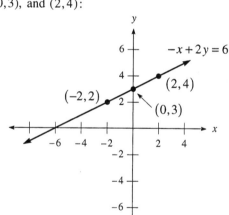

23. Three solutions are $(-4, 2)$, $(0, 0)$, and $(4, -2)$:

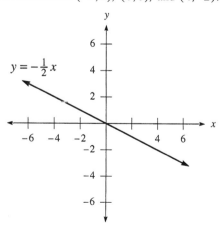

25. Three solutions are $(-1, -4)$, $(0, -1)$, and $(1, 2)$:

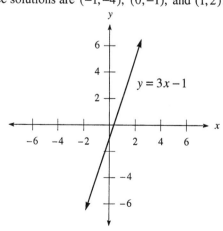

27.   Solving for y:

$$-2x + y = 1$$
$$y = 2x + 1$$

Three solutions are $(-2, -3)$, $(0, 1)$, and $(2, 5)$:

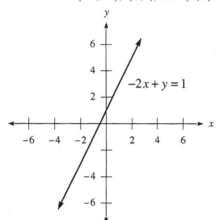

29.   Solving for y:

$$3x + 4y = 8$$
$$4y = -3x + 8$$
$$y = -\frac{3}{4}x + 2$$

Three solutions are $(-4, 5)$, $(0, 2)$, and $(4, -1)$:

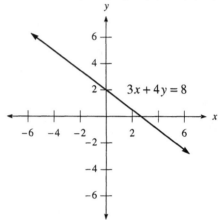

31.   Three solutions are $(-2, -4)$, $(-2, 0)$, and $(-2, 4)$:

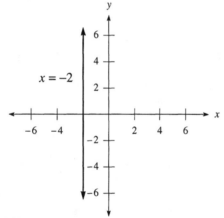

33.   Three solutions are $(-4, 2)$, $(0, 2)$, and $(4, 2)$:

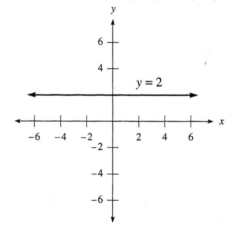

35.   Completing the table:

x	y
-4	-3
-2	-2
0	-1
2	0
6	2

37.   Graphing the three lines:

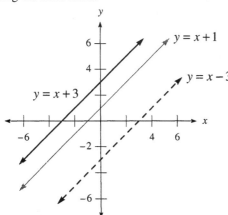

39.   The ordered pairs are $(1,4)$, $(2,3)$, and $(3,2)$:

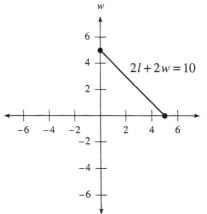

41.   Solving the equation:

$$3(x-2)=9$$
$$3x-6=9$$
$$3x=15$$
$$x=5$$

43.   Solving the equation:

$$2(3x-1)+4=-10$$
$$6x-2+4=-10$$
$$6x+2=-10$$
$$6x=-12$$
$$x=-2$$

45.   Solving the equation:

$$6-2(4x-7)=-4$$
$$6-8x+14=-4$$
$$-8x+20=-4$$
$$-8x=-24$$
$$x=3$$

47.   First multiply by 6 to clear the equation of fractions:

$$6\left(\tfrac{1}{2}x+4\right)=6\left(\tfrac{2}{3}x+5\right)$$
$$3x+24=4x+30$$
$$-x+24=30$$
$$-x=6$$
$$x=-6$$

3.4 More on Graphing: Intercepts

1.   To find the *x*-intercept, let $y = 0$:
$$2x+0=4$$
$$2x=4$$
$$x=2$$
To find the *y*-intercept, let $x = 0$:
$$2(0)+y=4$$
$$0+y=4$$
$$y=4$$
Graphing the line:

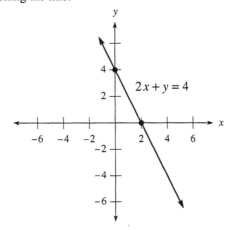

3.   To find the *x*-intercept, let $y = 0$:
$$-x+0=3$$
$$-x=3$$
$$x=-3$$
To find the *y*-intercept, let $x = 0$:
$$-0+y=3$$
$$y=3$$
Graphing the line:

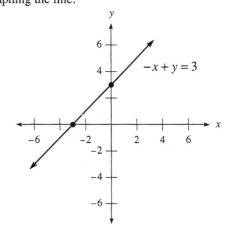

5. To find the x-intercept, let $y = 0$:
$$-x + 2(0) = 2$$
$$-x = 2$$
$$x = -2$$
To find the y-intercept, let $x = 0$:
$$-0 + 2y = 2$$
$$2y = 2$$
$$y = 1$$
Graphing the line:

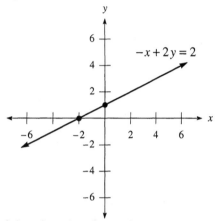

7. To find the x-intercept, let $y = 0$:
$$5x + 2(0) = 10$$
$$5x = 10$$
$$x = 2$$
To find the y-intercept, let $x = 0$:
$$5(0) + 2y = 10$$
$$2y = 10$$
$$y = 5$$
Graphing the line:

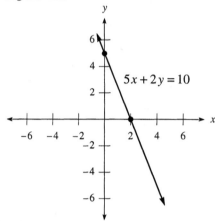

9. To find the x-intercept, let $y = 0$:
$$4x - 2(0) = 8$$
$$4x = 8$$
$$x = 2$$
To find the y-intercept, let $x = 0$:
$$4(0) - 2y = 8$$
$$-2y = 8$$
$$y = -4$$
Graphing the line:

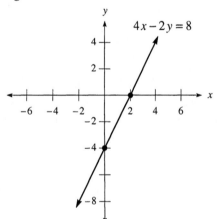

11. To find the x-intercept, let $y = 0$:
$$-4x + 5(0) = 20$$
$$-4x = 20$$
$$x = -5$$
To find the y-intercept, let $x = 0$:
$$-4(0) + 5y = 20$$
$$5y = 20$$
$$y = 4$$
Graphing the line:

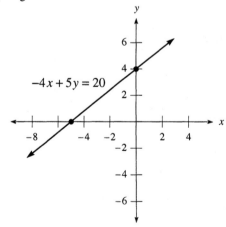

13. To find the x-intercept, let $y = 0$:
$$2x - 6 = 0$$
$$2x = 6$$
$$x = 3$$
To find the y-intercept, let $x = 0$:
$$y = 2(0) - 6$$
$$y = -6$$
Graphing the line:

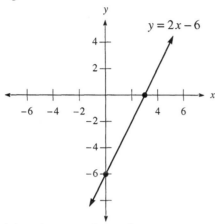

15. To find the x-intercept, let $y = 0$:
$$2x + 2 = 0$$
$$2x = -2$$
$$x = -1$$
To find the y-intercept, let $x = 0$:
$$y = 2(0) + 2$$
$$y = 2$$
Graphing the line:

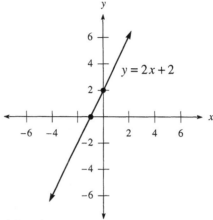

17. To find the x-intercept, let $y = 0$:
$$2x - 1 = 0$$
$$2x = 1$$
$$x = \frac{1}{2}$$
To find the y-intercept, let $x = 0$:
$$y = 2(0) - 1$$
$$y = -1$$
Graphing the line:

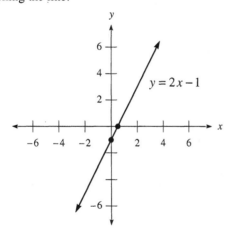

19. To find the x-intercept, let $y = 0$:
$$\tfrac{1}{2}x + 3 = 0$$
$$\tfrac{1}{2}x = -3$$
$$x = -6$$
To find the y-intercept, let $x = 0$:
$$y = \tfrac{1}{2}(0) + 3$$
$$y = 3$$
Graphing the line:

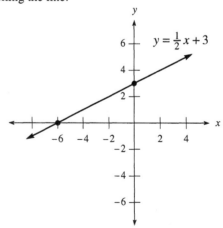

21.   To find the x-intercept, let $y = 0$:   To find the y-intercept, let $x = 0$:

$$-\tfrac{1}{3}x - 2 = 0$$
$$-\tfrac{1}{3}x = 2$$
$$x = -6$$

Graphing the line:

$$y = -\tfrac{1}{3}(0) - 2$$
$$y = -2$$

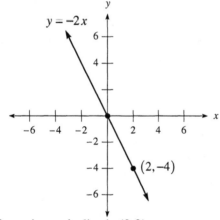

(graph for 21) $y = -\tfrac{1}{3}x - 2$

23.   Another point on the line is $(2,-4)$:

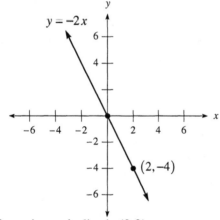

$y = -2x$ $(2,-4)$

25.   Another point on the line is $(3,-1)$:

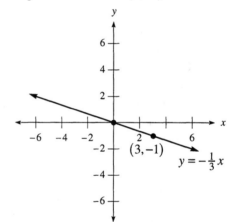

$(3,-1)$ $y = -\tfrac{1}{3}x$

27.   Another point on the line is $(3,2)$:

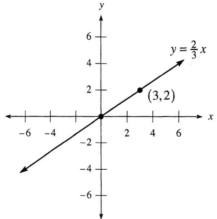

$y = \tfrac{2}{3}x$ $(3,2)$

29.   The x-intercept is 3 and the y-intercept is 5.   **31.**   The x-intercept is –1 and the y-intercept is –3.

33. Graphing the line:

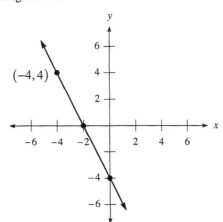

The *y*-intercept is –4.

35. Graphing the line:

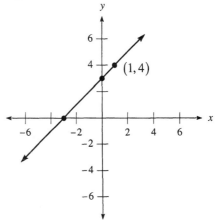

The *x*-intercept is –3.

37. Graphing the line:

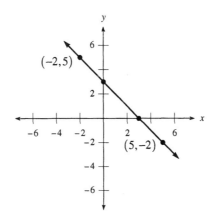

The *x*- and *y*-intercepts are both 3.

39. Completing the table:

x	y
–2	1
0	–1
–1	0
1	–2

41. Graphing the line:

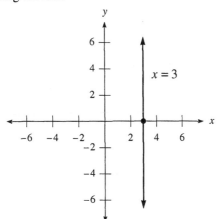

The *x*-intercept is 3.

43. Graphing the line:

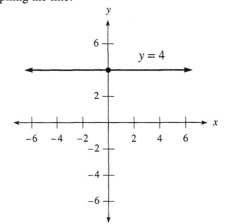

The *y*-intercept is 4.

45. Sketching the graph:

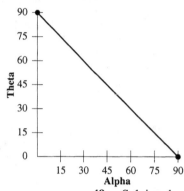

47. Solving the inequality:
$$x - 3 < 2$$
$$x - 3 + 3 < 2 + 3$$
$$x < 5$$

49. Solving the inequality:
$$-3x \geq 12$$
$$-\tfrac{1}{3}(-3x) \leq -\tfrac{1}{3}(12)$$
$$x \leq -4$$

51. Solving the inequality:
$$-\frac{x}{3} \leq -1$$
$$-3\left(-\frac{x}{3}\right) \geq -3(-1)$$
$$x \geq 3$$

53. Solving the inequality:
$$-4x + 1 < 17$$
$$-4x + 1 - 1 < 17 - 1$$
$$-4x < 16$$
$$-\tfrac{1}{4}(-4x) > -\tfrac{1}{4}(16)$$
$$x > -4$$

3.5 The Slope of a Line

1. The slope is given by: $m = \dfrac{4-1}{4-2} = \dfrac{3}{2}$

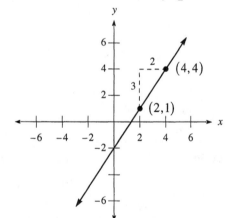

3. The slope is given by: $m = \dfrac{2-4}{5-1} = \dfrac{-2}{4} = -\dfrac{1}{2}$

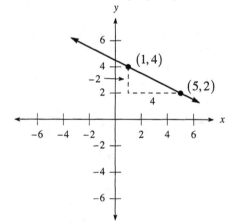

5.   The slope is given by: $m = \dfrac{2-(-3)}{4-1} = \dfrac{5}{3}$

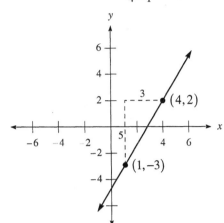

7.   The slope is given by: $m = \dfrac{3-(-2)}{1-(-3)} = \dfrac{5}{4}$

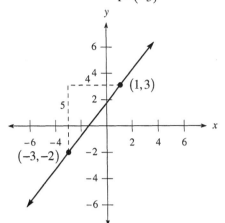

9.   The slope is given by: $m = \dfrac{-2-2}{3-(-3)} = \dfrac{-4}{6} = -\dfrac{2}{3}$

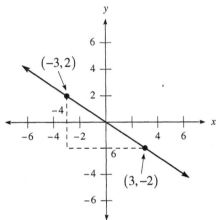

11.   The slope is given by: $m = \dfrac{-2-(-5)}{3-2} = \dfrac{3}{1} = 3$

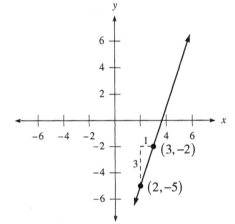

13.   Graphing the line:

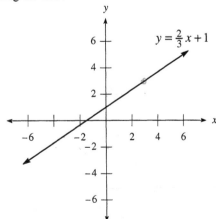

15.   Graphing the line:

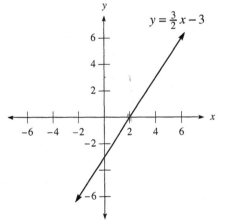

17.   Graphing the line:

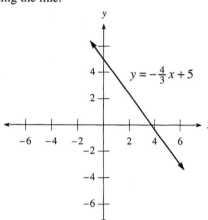

19.   Graphing the line:

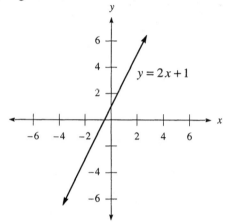

21.   Graphing the line:

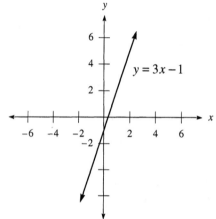

23.   The y-intercept is 2, and the slope is given by: $m = \dfrac{5-(-1)}{1-(-1)} = \dfrac{6}{2} = 3$

25.   The y-intercept is -2, and the slope is given by: $m = \dfrac{2-0}{2-1} = \dfrac{2}{1} = 2$

27.   The slope is given by: $m = \dfrac{0-(-2)}{3-0} = \dfrac{2}{3}$ **29.**   The slope is given by: $m = \dfrac{0-2}{4-0} = \dfrac{-2}{4} = -\dfrac{1}{2}$

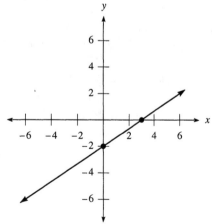

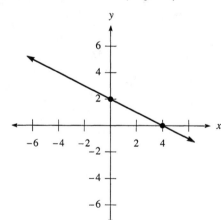

31.   Graphing the line:

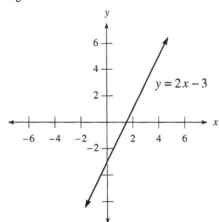

$y = 2x - 3$

The slope is 2 and the y-intercept is -3.

33.   Graphing the line:

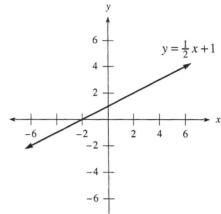

$y = \frac{1}{2}x + 1$

The slope is $\frac{1}{2}$ and the y-intercept is 1.

35.   Using the slope formula:

$$\frac{y-2}{6-4} = 2$$

$$\frac{y-2}{2} = 2$$

$$y - 2 = 4$$

$$y = 6$$

37.   Finding the slopes:

A: $\dfrac{121 - 88}{1970 - 1960} = \dfrac{33}{10} = 3.3$ B: $\dfrac{152 - 121}{1980 - 1970} = \dfrac{31}{10} = 3.1$

C: $\dfrac{205 - 152}{1990 - 1980} = \dfrac{53}{10} = 5.3$ D: $\dfrac{224 - 205}{2000 - 1990} = \dfrac{19}{10} = 1.9$

39.   Finding the slopes:

A: $\dfrac{28 - 33}{1999 - 1998} = \dfrac{-5}{1} = -5$ B: $\dfrac{23 - 25}{2001 - 2000} = \dfrac{-2}{1} = -2$ C: $\dfrac{20 - 22}{2003 - 2002} = \dfrac{-2}{1} = -2$

41.   Evaluating when $x = -3$: $2x - 9 = 2(-3) - 9 = -6 - 9 = -15$

43.   Evaluating when $x = -3$: $9 - 6x = 9 - 6(-3) = 9 + 18 = 27$

45.   Simplifying, then evaluating when $x = -3$: $4(3x + 2) + 1 = 12x + 8 + 1 = 12x + 9 = 12(-3) + 9 = -36 + 9 = -27$

47.   Evaluating when $x = -3$: $2x^2 + 3x + 4 = 2(-3)^2 + 3(-3) + 4 = 2(9) - 9 + 4 = 18 - 9 + 4 = 13$

3.6 Finding the Equation of a Line

1.   The slope-intercept form is $y = \frac{2}{3}x + 1$.

3.   The slope-intercept form is $y = \frac{3}{2}x - 1$.

5.   The slope-intercept form is $y = -\frac{2}{5}x + 3$.

7.   The slope-intercept form is $y = 2x - 4$.

9.   Solving for y:
$$-2x + y = 4$$
$$y = 2x + 4$$
The slope is 2 and the y-intercept is 4:

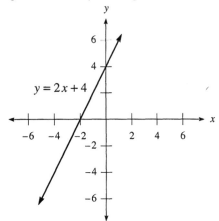

11.   Solving for y:
$$3x + y = 3$$
$$y = -3x + 3$$
The slope is –3 and the y-intercept is 3:

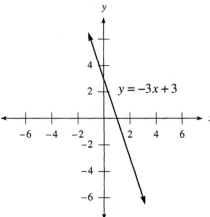

13.   Solving for y:
$$3x + 2y = 6$$
$$2y = -3x + 6$$
$$y = -\frac{3}{2}x + 3$$
The slope is $-\frac{3}{2}$ and the y-intercept is 3:

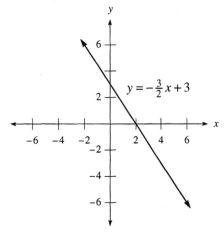

15.   Solving for y:
$$4x - 5y = 20$$
$$-5y = -4x + 20$$
$$y = \frac{4}{5}x - 4$$
The slope is $\frac{4}{5}$ and the y-intercept is –4:

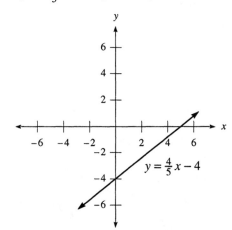

17.   Solving for y:
$$-2x - 5y = 10$$
$$-5y = 2x + 10$$
$$y = -\tfrac{2}{5}x - 2$$

The slope is $-\tfrac{2}{5}$ and the y-intercept is -2:

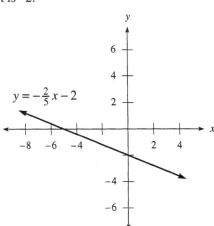

19.   Using the point-slope formula:
$$y - (-5) = 2(x - (-2))$$
$$y + 5 = 2(x + 2)$$
$$y + 5 = 2x + 4$$
$$y = 2x - 1$$

21.   Using the point-slope formula:
$$y - 1 = -\tfrac{1}{2}(x - (-4))$$
$$y - 1 = -\tfrac{1}{2}(x + 4)$$
$$y - 1 = -\tfrac{1}{2}x - 2$$
$$y = -\tfrac{1}{2}x - 1$$

23.   Using the point-slope formula:
$$y - (-3) = \tfrac{3}{2}(x - 2)$$
$$y + 3 = \tfrac{3}{2}x - 3$$
$$y = \tfrac{3}{2}x - 6$$

25.   Using the point-slope formula:
$$y - 4 = -3(x - (-1))$$
$$y - 4 = -3(x + 1)$$
$$y - 4 = -3x - 3$$
$$y = -3x + 1$$

27.   Finding the slope: $m = \dfrac{-1 - (-4)}{1 - (-2)} = \dfrac{-1 + 4}{1 + 2} = \dfrac{3}{3} = 1$

Using the point-slope formula:
$$y - (-4) = 1(x - (-2))$$
$$y + 4 = x + 2$$
$$y = x - 2$$

29.   Finding the slope: $m = \dfrac{1 - (-5)}{2 - (-1)} = \dfrac{1 + 5}{2 + 1} = \dfrac{6}{3} = 2$

Using the point-slope formula:
$$y - 1 = 2(x - 2)$$
$$y - 1 = 2x - 4$$
$$y = 2x - 3$$

31.   Finding the slope: $m = \dfrac{6 - (-2)}{3 - (-3)} = \dfrac{6 + 2}{3 + 3} = \dfrac{8}{6} = \dfrac{4}{3}$

Using the point-slope formula:
$$y - 6 = \tfrac{4}{3}(x - 3)$$
$$y - 6 = \tfrac{4}{3}x - 4$$
$$y = \tfrac{4}{3}x + 2$$

33.   Finding the slope: $m = \dfrac{-5 - (-1)}{3 - (-3)} = \dfrac{-5 + 1}{3 + 3} = \dfrac{-4}{6} = -\dfrac{2}{3}$

Using the point-slope formula:
$$y - (-5) = -\tfrac{2}{3}(x - 3)$$
$$y + 5 = -\tfrac{2}{3}x + 2$$
$$y = -\tfrac{2}{3}x - 3$$

35.   The y-intercept is 3, and the slope is: $m = \dfrac{3 - 0}{0 - (-1)} = \dfrac{3}{1} = 3$. The slope-intercept form is $y = 3x + 3$.

37.   The y-intercept is -1, and the slope is: $m = \dfrac{0 - (-1)}{4 - 0} = \dfrac{0 + 1}{4} = \dfrac{1}{4}$. The slope-intercept form is $y = \tfrac{1}{4}x - 1$.

39.   The slope is given by: $m = \dfrac{0 - 2}{3 - 0} = -\dfrac{2}{3}$. Since $b = 2$, the equation is $y = -\tfrac{2}{3}x + 2$.

41.   The slope is given by: $m = \dfrac{0 - (-5)}{-2 - 0} = -\dfrac{5}{2}$. Since $b = -5$, the equation is $y = -\tfrac{5}{2}x - 5$.

43.   Its equation is $x = 3$.

45. **a.** After 5 years, the copier is worth $6,000. **b.** The copier is worth $12,000 after 3 years.
  c. The slope of the line is –3000. **d.** The copier is decreasing in value by $3000 per year.
  e. The equation is $V = -3000t + 21,000$.

47. **a.** He will earn $3,000 for selling 1,000 shirts. **b.** He must sell 500 shirts to earn $2,000 for a month.
  c. The slope of the line is 2. **d.** Kevin earns $2 for each shirt he sells.
  e. The equation is $y = 2x + 1,000$.

49. Finding the amount:

$$(0.25)(300) = x$$
$$x = 75$$

So 75 is 25% of 300.

51. Finding the percent:
$$p \cdot 125 = 25$$
$$p = \frac{25}{125}$$
$$p = 0.2 = 20\%$$
So 25 is 20% of 125.

53. Finding the base:
$$0.15x = 60$$
$$x = \frac{60}{0.15}$$
$$x = 400$$
So 60 is 15% of 400.

55. Graphing the line:

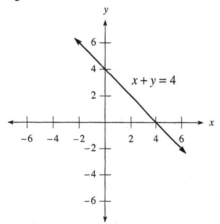

57. Graphing the line:

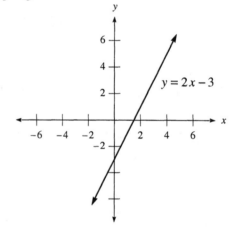

59. Graphing the line:

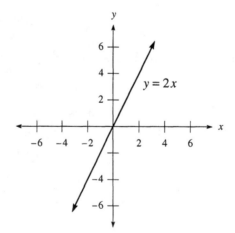

3.7 Linear Inequalities in Two Variables

1. Checking the point $(0,0)$:
$2(0)-3(0)=0-0<6$ (true)
Graphing the linear inequality:

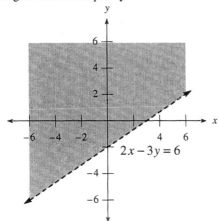

$2x-3y=6$

3. Checking the point $(0,0)$:
$0-2(0)=0-0\le 4$ (true)
Graphing the linear inequality:

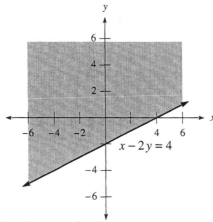

$x-2y=4$

5. Checking the point $(0,0)$:
$0-0\le 2$ (true)
Graphing the linear inequality:

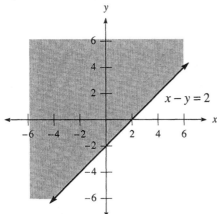

$x-y=2$

7. Checking the point $(0,0)$:
$3(0)-4(0)=0-0\ge 12$ (false)
Graphing the linear inequality:

$3x-4y=12$

9. Checking the point $(0,0)$:
$5(0)-0=0-0\le 5$ (true)
Graphing the linear inequality:

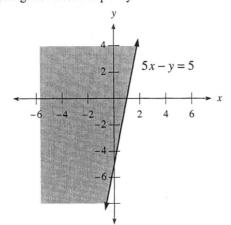

$5x-y=5$

11. Checking the point $(0,0)$:
$2(0)+6(0)=0+0\le 12$ (true)
Graphing the linear inequality:

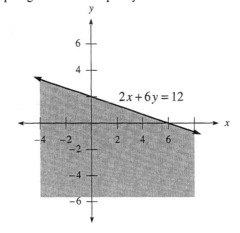

$2x+6y=12$

13.   Graphing the linear inequality:

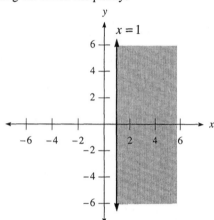

15.   Graphing the linear inequality:

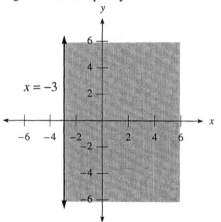

17.   Graphing the linear inequality:

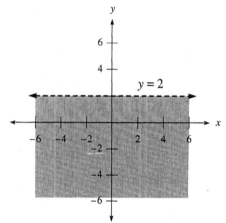

19.   Checking the point $(0,0)$:

$$2(0)+0 = 0+0 > 3 \qquad \text{(false)}$$

Graphing the linear inequality:

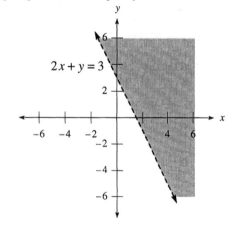

21.   Checking the point $(0,0)$:

$$0 \le 3(0)-1$$
$$0 \le -1 \qquad \text{(false)}$$

Graphing the linear inequality:

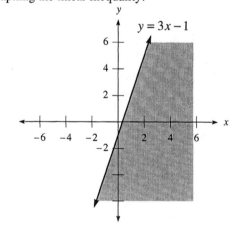

23.   Checking the point $(0,0)$:

$0 \le -\frac{1}{2}(0) + 2$

$0 \le 2$ (true)

Graphing the linear inequality:

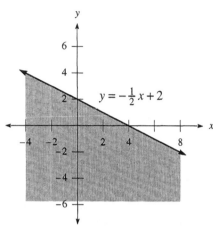

$y = -\frac{1}{2}x + 2$

25.   **a.**   The slope is $\frac{2}{5}$ and the y-intercept is 2, so the equation is $y = \frac{2}{5}x + 2$.

b.   Since the shading is below the line, the inequality is $y < \frac{2}{5}x + 2$.

c.   Since the shading is above the line, the inequality is $y > \frac{2}{5}x + 2$.

27.   Simplifying the expression: $7 - 3(2x - 4) - 8 = 7 - 6x + 12 - 8 = -6x + 11$

29.   Solving the equation:

$-\frac{3}{2}x = 12$

$-\frac{2}{3}\left(-\frac{3}{2}x\right) = -\frac{2}{3}(12)$

$x = -8$

31.   Solving the equation:

$8 - 2(x + 7) = 2$

$8 - 2x - 14 = 2$

$-2x - 6 = 2$

$-2x = 8$

$x = -4$

33.   Solving for w:

$2l + 2w = P$

$2w = P - 2l$

$w = \dfrac{P - 2l}{2}$

35.   Solving the inequality:

$3 - 2x > 5$

$3 - 3 - 2x > 5 - 3$

$-2x > 2$

$-\frac{1}{2}(-2x) < -\frac{1}{2}(2)$

$x < -1$

37.   Solving for y:

$3x - 2y \le 12$

$3x - 3x - 2y \le -3x + 12$

$-2y \le -3x + 12$

$-\frac{1}{2}(-2y) \ge -\frac{1}{2}(-3x + 12)$

$y \ge \frac{3}{2}x - 6$

Graphing the solution set:

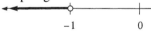

$\qquad -1 \qquad\qquad 0$

39.   Let w represent the width and $3w + 5$ represent the length. Using the perimeter formula:

$2(w) + 2(3w + 5) = 26$

$2w + 6w + 10 = 26$

$8w + 10 = 26$

$8w = 16$

$w = 2$

$3w + 5 = 3(2) + 5 = 11$

The width is 2 inches and the length is 11 inches.

Chapter 3 Review

1.   Substituting $x = 4$, $x = 0$, $y = 3$, and $y = 0$:

$$3(4) + y = 6 \qquad\qquad 3(0) + y = 6 \qquad\qquad 3x + 3 = 6 \qquad\qquad 3x + 0 = 6$$
$$12 + y = 6 \qquad\qquad\quad 0 + y = 6 \qquad\qquad\quad 3x = 3 \qquad\qquad\quad\; 3x = 6$$
$$y = -6 \qquad\qquad\qquad\; y = 6 \qquad\qquad\qquad x = 1 \qquad\qquad\qquad x = 2$$

The ordered pairs are $(4,-6), (0,6), (1,3)$, and $(2,0)$.

3.   Substituting $x = 4$, $y = -2$, and $y = 3$:

$$y = 2(4) - 6 \qquad\qquad -2 = 2x - 6 \qquad\qquad 3 = 2x - 6$$
$$y = 8 - 6 \qquad\qquad\quad\; 4 = 2x \qquad\qquad\qquad 9 = 2x$$
$$y = 2 \qquad\qquad\qquad\;\; x = 2 \qquad\qquad\qquad\; x = \frac{9}{2}$$

The ordered pairs are $(4,2), (2,-2)$, and $\left(\frac{9}{2}, 3\right)$.

5.   Substituting $x = 2$, $x = -1$, and $x = -3$ results (in each case) in $y = -3$. The ordered pairs are
$(2,-3), (-1,-3)$, and $(-3,-3)$.

7.   Substituting each ordered pair into the equation:

$$\left(-2, \tfrac{9}{2}\right): \quad 3(-2) - 4\left(\tfrac{9}{2}\right) = -6 - 18 = -24 \neq 12$$
$$(0,3): \quad 3(0) - 4(3) = 0 - 12 = -12 \neq 12$$
$$\left(2, -\tfrac{3}{2}\right): \quad 3(2) - 4\left(-\tfrac{3}{2}\right) = 6 + 6 = 12$$

Only the ordered pair $\left(2, -\frac{3}{2}\right)$ is a solution.

9.   Graphing the ordered pair:

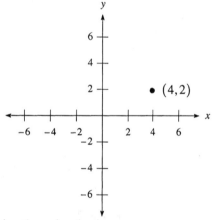

11. Graphing the ordered pair:

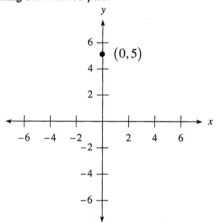

13. Graphing the ordered pair:

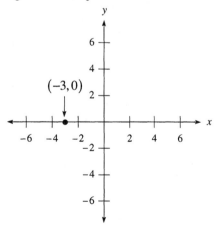

15. The ordered pairs are $(-2,0), (0,-2)$, and $(1,-3)$:

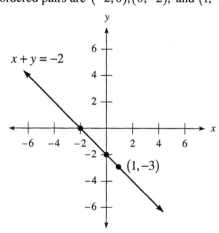

17. The ordered pairs are $(1,1),(0,-1)$, and $(-1,-3)$:

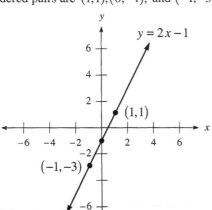

19. Graphing the equation:

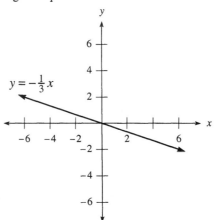

21. Graphing the equation:

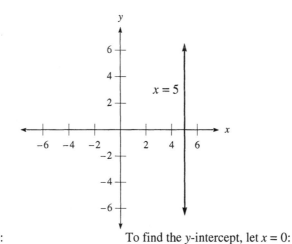

23. To find the x-intercept, let $y = 0$:

$$3x - 0 = 6$$
$$3x = 6$$
$$x = 2$$

To find the y-intercept, let $x = 0$:

$$3(0) - y = 6$$
$$-y = 6$$
$$y = -6$$

25. To find the x-intercept, let $y = 0$:

$$0 = x - 3$$
$$x = 3$$

To find the y-intercept, let $x = 0$:

$$y = 0 - 3$$
$$y = -3$$

27. The slope is given by: $m = \dfrac{5-3}{3-2} = \dfrac{2}{1} = 2$

29. The slope is given by: $m = \dfrac{-8-(-4)}{-3-(-1)} = \dfrac{-8+4}{-3+1} = \dfrac{-4}{-2} = 2$

31. Using the point-slope formula:

$$y - 4 = -2\big(x - (-1)\big)$$
$$y - 4 = -2(x + 1)$$
$$y - 4 = -2x - 2$$
$$y = -2x + 2$$

33. Using the point-slope formula:

$$y - (-2) = -\tfrac{3}{4}(x - 3)$$
$$y + 2 = -\tfrac{3}{4}x + \tfrac{9}{4}$$
$$y = -\tfrac{3}{4}x + \tfrac{1}{4}$$

35. The slope-intercept form is $y = -x + 6$.

37. The equation is in slope-intercept form, so $m = 4$ and $b = -1$.

39. Solving for y:

$$6x + 3y = 9$$
$$3y = -6x + 9$$
$$y = -2x + 3$$

The equation is in slope-intercept form, so $m = -2$ and $b = 3$.

41.   Checking the point $(0,0)$:

$$0 - 0 < 3 \quad \text{(true)}$$

Graphing the linear inequality:

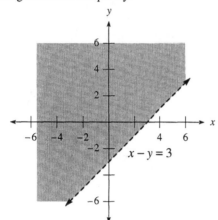

43.   Checking the point $(0,0)$:

$$0 \leq -2(0) + 3$$
$$0 \leq 3 \quad \text{(true)}$$

Graphing the linear inequality:

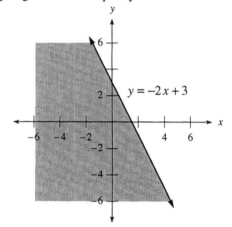

Chapters 1-3 Cumulative Review

1.   Simplifying using order of operations: $7 - 2 \cdot 6 = 7 - 12 = -5$

3.   Simplifying using order of operations: $4 \cdot 6 + 12 \div 4 - 3^2 = 24 + 3 - 9 = 18$

5.   Simplifying using order of operations: $(4 - 9)(-3 - 8) = (-5)(-11) = 55$

7.   First factor the denominators to find the LCM:

$$60 = 2 \cdot 2 \cdot 3 \cdot 5$$
$$84 = 2 \cdot 2 \cdot 3 \cdot 7$$
$$\text{LCM} = 2 \cdot 2 \cdot 3 \cdot 5 \cdot 7 = 420$$

Subtracting the fractions: $\frac{11}{60} - \frac{13}{84} = \frac{11 \cdot 7}{60 \cdot 7} - \frac{13 \cdot 5}{84 \cdot 5} = \frac{77}{420} - \frac{65}{420} = \frac{12}{420} = \frac{2 \cdot 2 \cdot 3}{2 \cdot 2 \cdot 3 \cdot 5 \cdot 7} = \frac{1}{5 \cdot 7} = \frac{1}{35}$

9.   Simplifying: $5a + 3 - 4a - 6 = 5a - 4a + 3 - 6 = a - 3$

11.   Solving the equation:

$$4x - 5 = 3$$
$$4x = 8$$
$$x = 2$$

13.   Solving the equation:

$$6(t + 5) - 4 = 2$$
$$6t + 30 - 4 = 2$$
$$6t + 26 = 2$$
$$6t = -24$$
$$t = -4$$

15.   Solving the equation:

$$0.05x + 0.07(200 - x) = 11$$
$$0.05x + 14 - 0.07x = 11$$
$$-0.02x + 14 = 11$$
$$-0.02x = -3$$
$$x = 150$$

17.   Solving the inequality:

$$5 - 7x \geq 19$$
$$-5 + 5 - 7x \geq -5 + 19$$
$$-7x \geq 14$$
$$-\tfrac{1}{7}(-7x) \leq -\tfrac{1}{7}(14)$$
$$x \leq -2$$

Graphing the solution set:

19.   Graphing the equation:

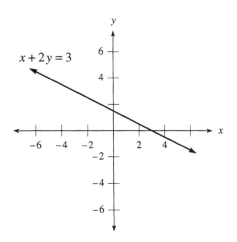

21.   Checking the point $(0,0)$:

$$0 - 2(0) \le 4$$
$$0 \le 4 \quad \text{(true)}$$

Graphing the linear inequality:

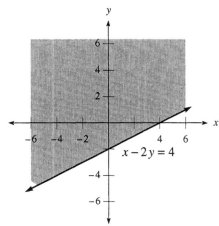

23.   Graphing the line which passes through $(-3,-2)$ and $(2,3)$:

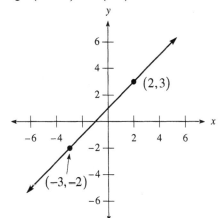

The point $(4,7)$ does not lie on the line, while $(1,2)$ does lie on the line.

25.   To find the x-intercept, let $y = 0$:   To find the y-intercept, let $x = 0$:

$$2x + 5(0) = 10 \qquad\qquad 2(0) + 5y = 10$$
$$2x = 10 \qquad\qquad\qquad 5y = 10$$
$$x = 5 \qquad\qquad\qquad\quad y = 2$$

Graphing the line:

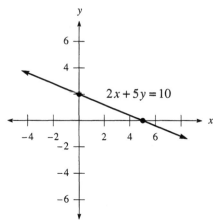

$$2x + 5y = 10$$

27. The slope is given by: $m = \dfrac{7-3}{5-2} = \dfrac{4}{3}$

29. Solving for y:
$$2x + 3y = 6$$
$$3y = -2x + 6$$
$$y = -\tfrac{2}{3}x + 2$$
The slope of the line is $-\tfrac{2}{3}$.

31. First find the slope: $m = \dfrac{7-3}{4-2} = \dfrac{4}{2} = 2$. Using the point-slope formula:
$$y - 3 = 2(x - 2)$$
$$y - 3 = 2x - 4$$
$$y = 2x - 1$$

33. Substituting $y = 3$ and $x = 0$:

$$3x - 2(3) = 6 \qquad\qquad\qquad 3(0) - 2y = 6$$
$$3x - 6 = 6 \qquad\qquad\qquad\qquad 0 - 2y = 6$$
$$3x = 12 \qquad\qquad\qquad\qquad\quad -2y = 6$$
$$x = 4 \qquad\qquad\qquad\qquad\qquad y = -3$$

The completed table is:

x	y
4	3
0	−3

35. Translating into symbols: $x + 7 = 4$

37. Each number in the sequence is the square of a fraction, since $\left(\tfrac{1}{2}\right)^2 = \tfrac{1}{4}$, $\left(\tfrac{1}{3}\right)^2 = \tfrac{1}{9}$, $\left(\tfrac{1}{4}\right)^2 = \tfrac{1}{16}$, and $\left(\tfrac{1}{5}\right)^2 = \tfrac{1}{25}$.

So the next number in the sequence is $\left(\tfrac{1}{6}\right)^2 = \tfrac{1}{36}$.

39. Computing the expression: $6(-2) - 5 = -12 - 5 = -17$

41. Reducing the fraction: $\dfrac{75}{135} = \dfrac{3 \cdot 5 \cdot 5}{3 \cdot 3 \cdot 3 \cdot 5} = \dfrac{5}{3 \cdot 3} = \dfrac{5}{9}$

43. Evaluating when $x = 2$: $x^2 + 6x - 7 = (2)^2 + 6(2) - 7 = 4 + 12 - 7 = 9$

45. Finding the percent:
$$p \cdot 36 = 27$$
$$p = \tfrac{27}{36}$$
$$p = 0.75 = 75\%$$
So 75% of 36 is 27.

47.   Let w represent the width and $2w + 5$ represent the length. Using the perimeter formula:

$$2(w) + 2(2w + 5) = 44$$
$$2w + 4w + 10 = 44$$
$$6w + 10 = 44$$
$$6w = 34$$
$$w = \frac{17}{3}$$
$$2w + 5 = 2\left(\frac{17}{3}\right) + 5 = \frac{34}{3} + 5 = \frac{49}{3}$$

The width is $\frac{17}{3}$ cm and the length is $\frac{49}{3}$ cm.

49.   Completing the table:

	Dollars Invested at 8%	Dollars Invested at 6%
Number of	$x + 900$	x
Interest on	$0.08(x + 900)$	$0.06(x)$

The equation is:

$$0.08(x + 900) + 0.06(x) = 240$$
$$0.08x + 72 + 0.06x = 240$$
$$0.14x + 72 = 240$$
$$0.14x = 168$$
$$x = 1200$$
$$x + 900 = 2100$$

Barbara invested $1,200 at 6% and $2,100 at 8%.

Chapter 3 Test

1.   Substituting $x = 0$, $y = 0$, $x = 10$, and $y = -3$:

$$2(0) - 5y = 10 \qquad 2x - 5(0) = 10 \qquad 2(10) - 5y = 10 \qquad 2x - 5(-3) = 10$$
$$0 - 5y = 10 \qquad 2x - 0 = 10 \qquad 20 - 5y = 10 \qquad 2x + 15 = 10$$
$$-5y = 10 \qquad 2x = 10 \qquad -5y = -10 \qquad 2x = -5$$
$$y = -2 \qquad x = 5 \qquad y = 2 \qquad x = -\frac{5}{2}$$

The ordered pairs are $(0, -2), (5, 0), (10, 2),$ and $\left(-\frac{5}{2}, -3\right)$.

2.   Substituting each ordered pair into the equation:

$$(2, 5): \qquad 4(2) - 3 = 8 - 3 = 5$$
$$(0, -3): \qquad 4(0) - 3 = 0 - 3 = -3$$
$$(3, 0): \qquad 4(3) - 3 = 12 - 3 = 9 \neq 0$$
$$(-2, 11): \qquad 4(-2) - 3 = -8 - 3 = -11 \neq 11$$

The ordered pairs $(2, 5)$ and $(0, -3)$ are solutions.

3.   Graphing the line:

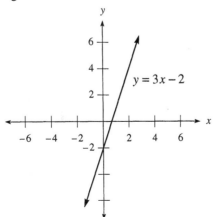

4.   Graphing the line:

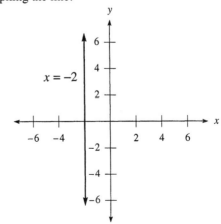

5. To find the x-intercept, let $y = 0$:
$$3x - 5(0) = 15$$
$$3x = 15$$
$$x = 5$$

To find the y-intercept, let $x = 0$:
$$3(0) - 5y = 15$$
$$-5y = 15$$
$$y = -3$$

6. To find the x-intercept, let $y = 0$:
$$0 = \frac{3}{2}x + 1$$
$$-1 = \frac{3}{2}x$$
$$x = -\frac{2}{3}$$

To find the y-intercept, let $x = 0$: $y = \frac{3}{2}(0) + 1 = 1$

7. The slope is given by: $m = \dfrac{-7 - (-3)}{4 - 2} = \dfrac{-4}{2} = -2$

8. The slope is given by: $m = \dfrac{-8 - 5}{2 - (-3)} = -\dfrac{13}{5}$

9. The slope is given by: $m = \dfrac{d - b}{c - a}$

10. The slope is given by: $m = \dfrac{4 - 3}{2x - 5x} = -\dfrac{1}{3x}$

11. Using the point-slope formula:
$$y - 5 = 3(x - (-2))$$
$$y - 5 = 3(x + 2)$$
$$y - 5 = 3x + 6$$
$$y = 3x + 11$$

12. The slope-intercept form is $y = 4x + 8$.

13. First find the slope: $m = \dfrac{4 - 1}{-2 - 3} = -\dfrac{3}{5}$. Using the point-slope formula:
$$y - 1 = -\frac{3}{5}(x - 3)$$
$$y - 1 = -\frac{3}{5}x + \frac{9}{5}$$
$$y = -\frac{3}{5}x + \frac{14}{5}$$

14. First find the slope: $m = \dfrac{4 - 0}{3 - 1} = \dfrac{4}{2} = 2$. Using the point-slope formula:
$$y - 4 = 2(x - 3)$$
$$y - 4 = 2x - 6$$
$$y = 2x - 2$$

15. Checking the point $(0,0)$:
$$0 < 0 + 4$$
$$0 < 4 \qquad \text{(true)}$$
Graphing the linear inequality:

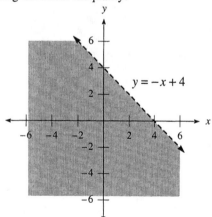

16. Checking the point $(0,0)$:
$$3(0) - 4(0) \geq 12$$
$$0 \geq 12 \qquad \text{(false)}$$
Graphing the linear inequality:

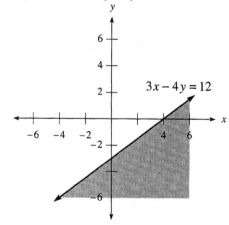

Chapter 4
Systems of Linear Equations

4.1 Solving Linear Systems by Graphing

1. Graphing both lines:

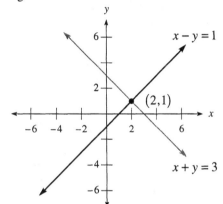

The intersection point is $(2,1)$.

3. Graphing both lines:

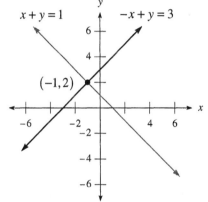

The intersection point is $(-1,2)$.

5. Graphing both lines:

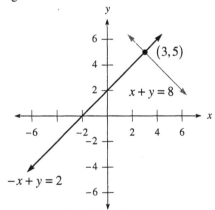

The intersection point is $(3,5)$.

7. Graphing both lines:

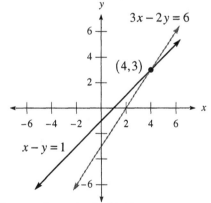

The intersection point is $(4,3)$.

9. Graphing both lines:

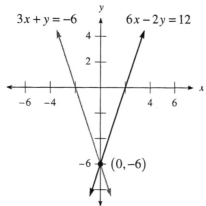

The intersection point is $(0, -6)$.

11. Graphing both lines:

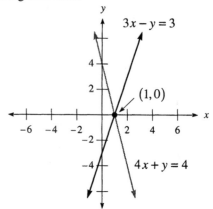

The intersection point is $(1, 0)$.

13. Graphing both lines:

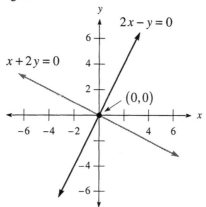

The intersection point is $(0, 0)$.

15. Graphing both lines:

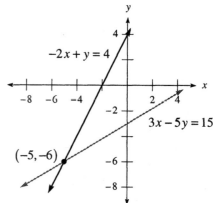

The intersection point is $(-5, -6)$.

17. Graphing both lines:

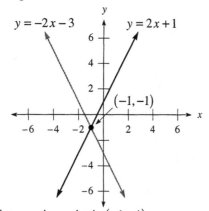

The intersection point is $(-1, -1)$.

19. Graphing both lines:

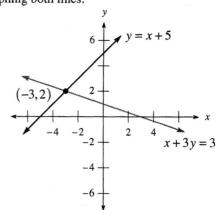

The intersection point is $(-3, 2)$.

21.   Graphing both lines:

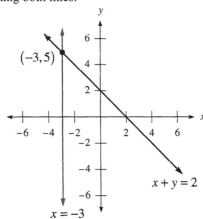

The intersection point is $(-3, 5)$.

23.   Graphing both lines:

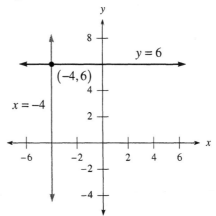

The intersection point is $(-4, 6)$.

25.   Graphing both lines:

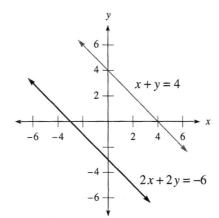

There is no intersection (the lines are parallel).

27.   Graphing both lines:

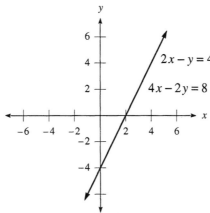

The system is dependent (both lines are the same, they coincide).

29. Graphing both lines:

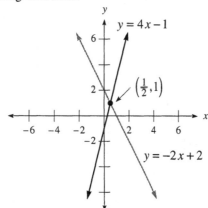

The intersection point is $\left(\frac{1}{2},1\right)$.

31. Graphing both lines:

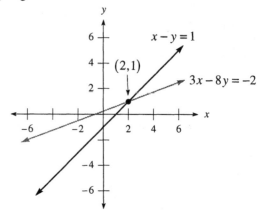

The intersection point is $(2,1)$.

33. **a.** If Jane worked 25 hours, she would earn the same amount at each position.
    b. If Jane worked less than 20 hours, she should choose Gigi's since she earns more in that position.
    c. If Jane worked more than 30 hours, she should choose Marcy's since she earns more in that position.

35. **a.** At 4 hours the cost of the two Internet providers will be the same.
    b. Computer Service will be the cheaper plan if Patrice will be on the Internet for less than 3 hours.
    c. ICM World will be the cheaper plan if Patrice will be on the Internet for more than 6 hours.

37. Finding the slope: $m = \dfrac{1-(-5)}{3-(-3)} = \dfrac{1+5}{3+3} = \dfrac{6}{6} = 1$

39. Finding the slope: $m = \dfrac{0-3}{5-0} = -\dfrac{3}{5}$

41. Graphing the line:

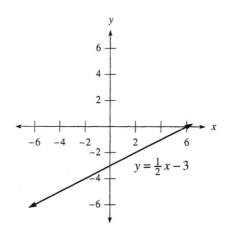

4.2 The Elimination Method

1. Adding the two equations:
$$2x = 4$$
$$x = 2$$
Substituting into the first equation:
$$2 + y = 3$$
$$y = 1$$
The solution is $(2,1)$.

3. Adding the two equations:
$$2y = 14$$
$$y = 7$$
Substituting into the first equation:
$$x + 7 = 10$$
$$x = 3$$
The solution is $(3,7)$.

5. Adding the two equations:
    $$-2y = 10$$
    $$y = -5$$
    Substituting into the first equation:
    $$x - (-5) = 7$$
    $$x + 5 = 7$$
    $$x = 2$$
    The solution is $(2, -5)$.

7. Adding the two equations:
    $$4x = -4$$
    $$x = -1$$
    Substituting into the first equation:
    $$-1 + y = -1$$
    $$y = 0$$
    The solution is $(-1, 0)$.

9. Adding the two equations:
    $$0 = 0$$
    The lines coincide (the system is dependent).

11. Multiplying the first equation by 2:
    $$6x - 2y = 8$$
    $$2x + 2y = 24$$
    Adding the two equations:
    $$8x = 32$$
    $$x = 4$$
    Substituting into the first equation:
    $$3(4) - y = 4$$
    $$12 - y = 4$$
    $$-y = -8$$
    $$y = 8$$
    The solution is $(4, 8)$.

13. Multiplying the second equation by -3:
    $$5x - 3y = -2$$
    $$-30x + 3y = -3$$
    Adding the two equations:
    $$-25x = -5$$
    $$x = \tfrac{1}{5}$$
    Substituting into the first equation:
    $$5\left(\tfrac{1}{5}\right) - 3y = -2$$
    $$1 - 3y = -2$$
    $$-3y = -3$$
    $$y = 1$$
    The solution is $\left(\tfrac{1}{5}, 1\right)$.

15. Multiplying the second equation by 4:
    $$11x - 4y = 11$$
    $$20x + 4y = 20$$
    Adding the two equations:
    $$31x = 31$$
    $$x = 1$$
    Substituting into the second equation:
    $$5(1) + y = 5$$
    $$5 + y = 5$$
    $$y = 0$$
    The solution is $(1, 0)$.

17. Multiplying the second equation by 3:
    $$3x - 5y = 7$$
    $$-3x + 3y = -3$$
    Adding the two equations:
    $$-2y = 4$$
    $$y = -2$$
    Substituting into the second equation:
    $$-x - 2 = -1$$
    $$-x = 1$$
    $$x = -1$$
    The solution is $(-1, -2)$.

19. Multiplying the first equation by -2:
    $$2x + 16y = 2$$
    $$-2x + 4y = 13$$
    Adding the two equations:
    $$20y = 15$$
    $$y = \tfrac{3}{4}$$
    Substituting into the first equation:
    $$-x - 8\left(\tfrac{3}{4}\right) = -1$$
    $$-x - 6 = -1$$
    $$-x = 5$$
    $$x = -5$$
    The solution is $\left(-5, \tfrac{3}{4}\right)$.

21. Multiplying the first equation by 2:
    $$-6x - 2y = 14$$
    $$6x + 7y = 11$$
    Adding the two equations:
    $$5y = 25$$
    $$y = 5$$
    Substituting into the first equation:
    $$-3x - 5 = 7$$
    $$-3x = 12$$
    $$x = -4$$
    The solution is $(-4, 5)$.

23.   Adding the two equations:
$$8x = -24$$
$$x = -3$$
Substituting into the second equation:
$$2(-3) + y = -16$$
$$-6 + y = -16$$
$$y = -10$$
The solution is $(-3, -10)$.

25.   Multiplying the second equation by 3:
$$x + 3y = 9$$
$$6x - 3y = 12$$
Adding the two equations:
$$7x = 21$$
$$x = 3$$
Substituting into the first equation:

$$3 + 3y = 9$$
$$3y = 6$$
$$y = 2$$

The solution is $(3, 2)$.

27.   Multiplying the second equation by 2:
$$x - 6y = 3$$
$$8x + 6y = 42$$
Adding the two equations:
$$9x = 45$$
$$x = 5$$
Substituting into the second equation:
$$4(5) + 3y = 21$$
$$20 + 3y = 21$$
$$3y = 1$$
$$y = \frac{1}{3}$$
The solution is $\left(5, \frac{1}{3}\right)$.

29.   Multiplying the second equation by –3:
$$2x + 9y = 2$$
$$-15x - 9y = 24$$
Adding the two equations:
$$-13x = 26$$
$$x = -2$$
Substituting into the first equation:
$$2(-2) + 9y = 2$$
$$-4 + 9y = 2$$
$$9y = 6$$
$$y = \frac{2}{3}$$
The solution is $\left(-2, \frac{2}{3}\right)$.

31.   To clear each equation of fractions, multiply the first equation by 12 and the second equation by 6:
$$12\left(\tfrac{1}{3}x + \tfrac{1}{4}y\right) = 12\left(\tfrac{7}{6}\right) \qquad\qquad 6\left(\tfrac{3}{2}x - \tfrac{1}{3}y\right) = 6\left(\tfrac{7}{3}\right)$$
$$4x + 3y = 14 \qquad\qquad\qquad 9x - 2y = 14$$
The system of equations is:
$$4x + 3y = 14$$
$$9x - 2y = 14$$
Multiplying the first equation by 2 and the second equation by 3:
$$8x + 6y = 28$$
$$27x - 6y = 42$$
Adding the two equations:
$$35x = 70$$
$$x = 2$$
Substituting into $4x + 3y = 14$:
$$4(2) + 3y = 14$$
$$8 + 3y = 14$$
$$3y = 6$$
$$y = 2$$
The solution is $(2, 2)$.

33.   Multiplying the first equation by –2:
$$-6x - 4y = 2$$
$$6x + 4y = 0$$
Adding the two equations:
$$0 = 2$$
Since this statement is false, the two lines are parallel, so the system has no solution.

35.   Multiplying the first equation by 2 and the second equation by 3:
$$22x + 12y = 34$$
$$15x - 12y = 3$$
Adding the two equations:
$$37x = 37$$
$$x = 1$$
Substituting into the second equation:
$$5(1) - 4y = 1$$
$$5 - 4y = 1$$
$$-4y = -4$$
$$y = 1$$
The solution is $(1,1)$.

37.   To clear each equation of fractions, multiply the first equation by 6 and the second equation by 6:
$$6\left(\tfrac{1}{2}x + \tfrac{1}{6}y\right) = 6\left(\tfrac{1}{3}\right) \qquad\qquad 6\left(-x - \tfrac{1}{3}y\right) = 6\left(-\tfrac{1}{6}\right)$$
$$3x + y = 2 \qquad\qquad\qquad\qquad -6x - 2y = -1$$
The system of equations is:
$$3x + y = 2$$
$$-6x - 2y = -1$$
Multiplying the first equation by 2:
$$6x + 2y = 4$$
$$-6x - 2y = -1$$
Adding the two equations:
$$0 = 3$$
Since this statement is false, the two lines are parallel, so the system has no solution.

39.   Adding $2x$ to each side of the first equation and $-3x$ to each side of the second equation:
$$6x - 5y = 17$$
$$-3x + 5y = 4$$
Adding the two equations:
$$3x = 21$$
$$x = 7$$
Substituting into $5y = 3x + 4$:
$$5y = 3(7) + 4$$
$$5y = 21 + 4$$
$$5y = 25$$
$$y = 5$$
The solution is $(7,5)$.

41.   Multiplying the second equation by 100 (to eliminate decimals):
$$x + y = 22$$
$$5x + 10y = 170$$
Multiplying the first equation by –5:
$$-5x - 5y = -110$$
$$5x + 10y = 170$$
Adding the two equations:
$$5y = 60$$
$$y = 12$$
Substituting into the first equation:
$$x + 12 = 22$$
$$x = 10$$
The solution is $(10,12)$.

43. Solving for y:

$$-2x + 4y = 8$$
$$4y = 2x + 8$$
$$y = \tfrac{1}{2}x + 2$$

So $m = \tfrac{1}{2}$ and $b = 2$.

47. First find the slope: $m = \dfrac{1-(-5)}{3-(-3)} = \dfrac{1+5}{3+3} = \dfrac{6}{6} = 1$

Using the point-slope formula:
$$y - 1 = 1(x - 3)$$
$$y - 1 = x - 3$$
$$y = x - 2$$

45. Using the point-slope formula:
$$y - (-6) = 3(x - (-2))$$
$$y + 6 = 3(x + 2)$$
$$y + 6 = 3x + 6$$
$$y = 3x$$

4.3 The Substitution Method

1. Substituting into the first equation:
$$x + (2x - 1) = 11$$
$$3x - 1 = 11$$
$$3x = 12$$
$$x = 4$$
$$y = 2(4) - 1 = 7$$
The solution is $(4, 7)$.

5. Substituting into the first equation:
$$-2x + (-4x + 8) = -1$$
$$-6x + 8 = -1$$
$$-6x = -9$$
$$x = \tfrac{3}{2}$$
$$y = -4\left(\tfrac{3}{2}\right) + 8 = -6 + 8 = 2$$
The solution is $\left(\tfrac{3}{2}, 2\right)$.

9. Substituting into the first equation:
$$5x - 4(4) = -16$$
$$5x - 16 = -16$$
$$5x = 0$$
$$x = 0$$
The solution is $(0, 4)$.

13. Solving the second equation for x:
$$x - 2y = -1$$
$$x = 2y - 1$$
Substituting into the first equation:
$$(2y - 1) + 3y = 4$$
$$5y - 1 = 4$$
$$5y = 5$$
$$y = 1$$
$$x = 2(1) - 1 = 1$$
The solution is $(1, 1)$.

2. Substituting into the first equation:
$$x + (5x + 2) = 20$$
$$6x + 2 = 20$$
$$6x = 18$$
$$x = 3$$
$$y = 5(3) + 2 = 17$$
The solution is $(3, 17)$.

7. Substituting into the first equation:
$$3(-y + 6) - 2y = -2$$
$$-3y + 18 - 2y = -2$$
$$-5y + 18 = -2$$
$$-5y = -20$$
$$y = 4$$
$$x = -4 + 6 = 2$$
The solution is $(2, 4)$.

11. Substituting into the first equation:
$$5x + 4(-3x) = 7$$
$$5x - 12x = 7$$
$$-7x = 7$$
$$x = -1$$
$$y = -3(-1) = 3$$
The solution is $(-1, 3)$.

15. Solving the first equation for x:
$$x - 5y = 17$$
$$x = 5y + 17$$
Substituting into the first equation:
$$2(5y + 17) + y = 1$$
$$10y + 34 + y = 1$$
$$11y + 34 = 1$$
$$11y = -33$$
$$y = -3$$
$$x = 5(-3) + 17 = 2$$
The solution is $(2, -3)$.

17. Solving the second equation for x:
$$x - 5y = -5$$
$$x = 5y - 5$$
Substituting into the first equation:
$$3(5y - 5) + 5y = -3$$
$$15y - 15 + 5y = -3$$
$$20y - 15 = -3$$
$$20y = 12$$
$$y = \frac{3}{5}$$
$$x = 5\left(\frac{3}{5}\right) - 5 = 3 - 5 = -2$$
The solution is $\left(-2, \frac{3}{5}\right)$.

19. Solving the second equation for x:
$$x - 3y = -18$$
$$x = 3y - 18$$
Substituting into the first equation:
$$5(3y - 18) + 3y = 0$$
$$15y - 90 + 3y = 0$$
$$18y - 90 = 0$$
$$18y = 90$$
$$y = 5$$
$$x = 3(5) - 18 = -3$$
The solution is $(-3, 5)$.

21. Solving the second equation for x:
$$x + 3y = 12$$
$$x = -3y + 12$$
Substituting into the first equation:
$$-3(-3y + 12) - 9y = 7$$
$$9y - 36 - 9y = 7$$
$$-36 = 7$$
Since this statement is false, there is no solution to the system. The two lines are parallel.

23. Substituting into the first equation:
$$5x - 8(2x - 5) = 7$$
$$5x - 16x + 40 = 7$$
$$-11x + 40 = 7$$
$$-11x = -33$$
$$x = 3$$
$$y = 2(3) - 5 = 1$$
The solution is $(3, 1)$.

25. Substituting into the first equation:
$$7(2y - 1) - 6y = -1$$
$$14y - 7 - 6y = -1$$
$$8y - 7 = -1$$
$$8y = 6$$
$$y = \frac{3}{4}$$
$$x = 2\left(\frac{3}{4}\right) - 1 = \frac{3}{2} - 1 = \frac{1}{2}$$
The solution is $\left(\frac{1}{2}, \frac{3}{4}\right)$.

27. Substituting into the first equation:
$$-3x + 2(3x) = 6$$
$$-3x + 6x = 6$$
$$3x = 6$$
$$x = 2$$
$$y = 3(2) = 6$$
The solution is $(2, 6)$.

29. Substituting into the first equation:
$$5(y) - 6y = -4$$
$$-y = -4$$
$$y = 4$$
$$x = 4$$
The solution is $(4, 4)$.

31. Substituting into the first equation:
$$3x + 3(2x - 12) = 9$$
$$3x + 6x - 36 = 9$$
$$9x - 36 = 9$$
$$9x = 45$$
$$x = 5$$
$$y = 2(5) - 12 = -2$$
The solution is $(5, -2)$.

33. Substituting into the first equation:
$$7x - 11(10) = 16$$
$$7x - 110 = 16$$
$$7x = 126$$
$$x = 18$$
$$y = 10$$
The solution is $(18, 10)$.

35. Substituting into the first equation:
$$-4x + 4(x - 2) = -8$$
$$-4x + 4x - 8 = -8$$
$$-8 = -8$$
Since this statement is true, the system is dependent. The two lines coincide.

37.   Substituting into the first equation:
$$0.05x + 0.10(22 - x) = 1.70$$
$$0.05x + 2.2 - 0.10x = 1.70$$
$$-0.05x + 2.2 = 1.7$$
$$-0.05x = -0.5$$
$$x = 10$$
$$y = 22 - 10 = 12$$
The solution is $(10, 12)$.

39.   a.   At 1,000 miles the car and truck cost the same to operate.
      b.   If Daniel drives more than 1,200 miles, the car will be cheaper to operate.
      c.   If Daniel drives less than 800 miles, the truck will be cheaper to operate.
      d.   The graphs appear in the first quadrant only because all quantities are positive.

41.   Let x represent the number of gallons consumed. For the charges for each company to be the same, the equation is:
$$7.00 + 1.10x = 5.00 + 1.15x$$
$$7.00 - 0.05x = 5.00$$
$$-0.05x = -2$$
$$x = 40$$
If 40 gallons are used in a month, the two companies charge the same amount.

43.   Let w represent the width, and $3w$ represent the length. Using the perimeter formula:
$$2(w) + 2(3w) = 24$$
$$2w + 6w = 24$$
$$8w = 24$$
$$w = 3$$
$$3w = 9$$
The length is 9 meters and the width is 3 meters.

45.   Completing the table:

	Nickels	Dimes
Number	x	$x + 3$
Value (cents)	$5(x)$	$10(x+3)$

The equation is:
$$5(x) + 10(x + 3) = 210$$
$$5x + 10x + 30 = 210$$
$$15x + 30 = 210$$
$$15x = 180$$
$$x = 12$$
$$x + 3 = 15$$
The collection consists of 12 nickels and 15 dimes.

47.   Finding the amount:
$$0.08(6000) = x$$
$$x = 480$$
So 8% of 6,000 is 480.

49.   Completing the table:

	Dollars Invested at 8%	Dollars Invested at 10%
Number of	x	$2x$
Interest on	$0.08(x)$	$0.10(2x)$

The equation is:
$$0.08(x) + 0.10(2x) = 224$$
$$0.08x + 0.20x = 224$$
$$0.28x = 224$$
$$x = 800$$
$$2x = 1600$$
The man invested $800 at 8% interest and $1,600 at 10% interest.

4.4 Applications

1. Let x and y represent the two numbers. The system of equations is:

$$x + y = 25$$
$$y = 5 + x$$

Substituting into the first equation:

$$x + (5 + x) = 25$$
$$2x + 5 = 25$$
$$2x = 20$$
$$x = 10$$
$$y = 5 + 10 = 15$$

The two numbers are 10 and 15.

3. Let x and y represent the two numbers. The system of equations is:

$$x + y = 15$$
$$y = 4x$$

Substituting into the first equation:

$$x + 4x = 15$$
$$5x = 15$$
$$x = 3$$
$$y = 4(3) = 12$$

The two numbers are 3 and 12.

5. Let x represent the larger number and y represent the smaller number. The system of equations is:

$$x - y = 5$$
$$x = 2y + 1$$

Substituting into the first equation:

$$2y + 1 - y = 5$$
$$y + 1 = 5$$
$$y = 4$$
$$x = 2(4) + 1 = 9$$

The two numbers are 4 and 9.

7. Let x and y represent the two numbers. The system of equations is:

$$y = 4x + 5$$
$$x + y = 35$$

Substituting into the second equation:

$$x + 4x + 5 = 35$$
$$5x + 5 = 35$$
$$5x = 30$$
$$x = 6$$
$$y = 4(6) + 5 = 29$$

The two numbers are 6 and 29.

9. Let x represent the amount invested at 6% and y represent the amount invested at 8%. The system of equations is:

$$x + y = 20000$$
$$0.06x + 0.08y = 1380$$

Multiplying the first equation by –0.06:

$$-0.06x - 0.06y = -1200$$
$$0.06x + 0.08y = 1380$$

Adding the two equations:

$$0.02y = 180$$
$$y = 9000$$

Substituting into the first equation:

$$x + 9000 = 20000$$
$$x = 11000$$

Mr. Wilson invested $9,000 at 8% and $11,000 at 6%.

11. Let x represent the amount invested at 5% and y represent the amount invested at 6%. The system of equations is:
$$x = 4y$$
$$0.05x + 0.06y = 520$$
Substituting into the second equation:
$$0.05(4y) + 0.06y = 520$$
$$0.20y + 0.06y = 520$$
$$0.26y = 520$$
$$y = 2000$$
$$x = 4(2000) = 8000$$
She invested $8,000 at 5% and $2,000 at 6%.

13. Let x represent the number of nickels and y represent the number of quarters. The system of equations is:
$$x + y = 14$$
$$0.05x + 0.25y = 2.30$$
Multiplying the first equation by –0.05:
$$-0.05x - 0.05y = -0.7$$
$$0.05x + 0.25y = 2.30$$
Adding the two equations:
$$0.20y = 1.6$$
$$y = 8$$
Substituting into the first equation:
$$x + 8 = 14$$
$$x = 6$$
Ron has 6 nickels and 8 quarters.

15. Let x represent the number of dimes and y represent the number of quarters. The system of equations is:
$$x + y = 21$$
$$0.10x + 0.25y = 3.45$$
Multiplying the first equation by –0.10:
$$-0.10x - 0.10y = -2.10$$
$$0.10x + 0.25y = 3.45$$
Adding the two equations:
$$0.15y = 1.35$$
$$y = 9$$
Substituting into the first equation:
$$x + 9 = 21$$
$$x = 12$$
Tom has 12 dimes and 9 quarters.

17. Let x represent the liters of 50% alcohol solution and y represent the liters of 20% alcohol solution. The system of equations is:
$$x + y = 18$$
$$0.50x + 0.20y = 0.30(18)$$
Multiplying the first equation by –0.20:
$$-0.20x - 0.20y = -3.6$$
$$0.50x + 0.20y = 5.4$$
Adding the two equations:
$$0.30x = 1.8$$
$$x = 6$$
Substituting into the first equation:
$$6 + y = 18$$
$$y = 12$$
The mixture contains 6 liters of 50% alcohol solution and 12 liters of 20% alcohol solution.

19. Let x represent the gallons of 10% disinfectant and y represent the gallons of 7% disinfectant. The system of equations is:

$$x + y = 30$$
$$0.10x + 0.07y = 0.08(30)$$

Multiplying the first equation by -0.07:

$$-0.07x - 0.07y = -2.1$$
$$0.10x + 0.07y = 2.4$$

Adding the two equations:

$$0.03x = 0.3$$
$$x = 10$$

Substituting into the first equation:

$$10 + y = 30$$
$$y = 20$$

The mixture contains 10 gallons of 10% disinfectant and 20 gallons of 7% disinfectant.

21. Let x represent the number of adult tickets and y represent the number of kids tickets. The system of equations is:

$$x + y = 70$$
$$5.50x + 4.00y = 310$$

Multiplying the first equation by -4:

$$-4.00x - 4.00y = -280$$
$$5.50x + 4.00y = 310$$

Adding the two equations:

$$1.5x = 30$$
$$x = 20$$

Substituting into the first equation:

$$20 + y = 70$$
$$y = 50$$

The matinee had 20 adult tickets sold and 50 kids tickets sold.

23. Let x represent the width and y represent the length. The system of equations is:

$$2x + 2y = 96$$
$$y = 2x$$

Substituting into the first equation:

$$2x + 2(2x) = 96$$
$$2x + 4x = 96$$
$$6x = 96$$
$$x = 16$$
$$y = 2(16) = 32$$

The width is 16 feet and the length is 32 feet.

25. Let x represent the number of $5 chips and y represent the number of $25 chips. The system of equations is:

$$x + y = 45$$
$$5x + 25y = 465$$

Multiplying the first equation by -5:

$$-5x - 5y = -225$$
$$5x + 25y = 465$$

Adding the two equations:

$$20y = 240$$
$$y = 12$$

Substituting into the first equation:

$$x + 12 = 45$$
$$x = 33$$

The gambler has 33 $5 chips and 12 $25 chips.

27. Let x represent the number of shares of \$11 stock and y represent the number of shares of \$20 stock. The system of equations is:
$$x + y = 150$$
$$11x + 20y = 2550$$
Multiplying the first equation by -11:
$$-11x - 11y = -1650$$
$$11x + 20y = 2550$$
Adding the two equations:
$$9y = 900$$
$$y = 100$$
Substituting into the first equation:
$$x + 100 = 150$$
$$x = 50$$
She bought 50 shares at \$11 and 100 shares at \$20.

29. Substituting $x = -2$, $x = 0$, and $x = 2$:
$$y = \tfrac{1}{2}(-2) + 3 = -1 + 3 = 2 \qquad y = \tfrac{1}{2}(0) + 3 = 0 + 3 = 3 \qquad y = \tfrac{1}{2}(2) + 3 = 1 + 3 = 4$$
The ordered pairs are $(-2, 2)$, $(0, 3)$, and $(2, 4)$.

31. Graphing the line:

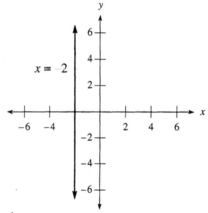

33. Computing the slope: $m = \dfrac{1-5}{0-2} = \dfrac{-4}{-2} = 2$

35. Using the point-slope formula:
$$y - 1 = \tfrac{1}{2}(x - (-2))$$
$$y - 1 = \tfrac{1}{2}(x + 2)$$
$$y - 1 = \tfrac{1}{2}x + 1$$
$$y = \tfrac{1}{2}x + 2$$

37. Computing the slope: $m = \dfrac{1-5}{0-2} = \dfrac{-4}{-2} = 2$. Using the point-slope formula:
$$y - 1 = 2(x - 0)$$
$$y - 1 = 2x$$
$$y = 2x + 1$$

Chapter 4 Review

1. Graphing both lines:

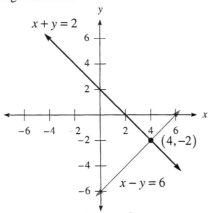

The intersection point is $(4,-2)$.

3. Graphing both lines:

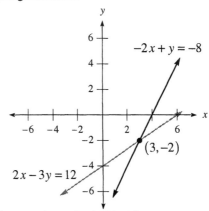

The intersection point is $(3,-2)$.

5. Graphing both lines:

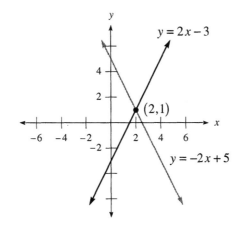

The intersection point is $(2,1)$.

7. Adding the two equations:
$$2x = 2$$
$$x = 1$$
Substituting into the second equation:
$$1 + y = -2$$
$$y = -3$$
The solution is $(1,-3)$.

9. Multiplying the first equation by 2:
$$10x - 6y = 4$$
$$-10x + 6y = -4$$
Adding the two equations:
$$0 = 0$$
Since this statement is true, the system is dependent. The two lines coincide.

11. Multiplying the second equation by –4:
$$-3x + 4y = 1$$
$$16x - 4y = 12$$
Adding the two equations:
$$13x = 13$$
$$x = 1$$
Substituting into the second equation:
$$-4(1) + y = -3$$
$$-4 + y = -3$$
$$y = 1$$
The solution is $(1,1)$.

13. Multiplying the first equation by 3 and the second equation by 5:
$$-6x + 15y = -33$$
$$35x - 15y = -25$$
Adding the two equations:
$$29x = -58$$
$$x = -2$$
Substituting into the first equation:
$$-2(-2) + 5y = -11$$
$$4 + 5y = -11$$
$$5y = -15$$
$$y = -3$$
The solution is $(-2,-3)$.

15. Substituting into the first equation:
$$x + (-3x + 1) = 5$$
$$-2x + 1 = 5$$
$$-2x = 4$$
$$x = -2$$
Substituting into the second equation: $y = -3(-2) + 1 = 6 + 1 = 7$. The solution is $(-2,7)$.

17. Substituting into the first equation:
$$4x - 3(3x + 7) = -16$$
$$4x - 9x - 21 = -16$$
$$-5x - 21 = -16$$
$$-5x = 5$$
$$x = -1$$
Substituting into the second equation: $y = 3(-1) + 7 = -3 + 7 = 4$. The solution is $(-1,4)$.

19. Solving the first equation for x:
$$x - 4y = 2$$
$$x = 4y + 2$$
Substituting into the second equation:
$$-3(4y + 2) + 12y = -8$$
$$-12y - 6 + 12y = -8$$
$$-6 = -8$$
Since this statement is false, there is no solution to the system. The two lines are parallel.

21. Solving the second equation for x:
$$x + 6y = -11$$
$$x = -6y - 11$$
Substituting into the first equation:
$$10(-6y - 11) - 5y = 20$$
$$-60y - 110 - 5y = 20$$
$$-65y - 110 = 20$$
$$-65y = 130$$
$$y = -2$$
Substituting into $x = -6y - 11$: $x = -6(-2) - 11 = 12 - 11 = 1$. The solution is $(1,-2)$.

23. Let x represent the smaller number and y represent the larger number. The system of equations is:
$$x + y = 18$$
$$2x = 6 + y$$
Solving the first equation for y:
$$x + y = 18$$
$$y = -x + 18$$
Substituting into the second equation:
$$2x = 6 + (-x + 18)$$
$$2x = -x + 24$$
$$3x = 24$$
$$x = 8$$
Substituting into the first equation:
$$8 + y = 18$$
$$y = 10$$
The two numbers are 8 and 10.

25. Let x represent the amount invested at 4% and y represent the amount invested at 5%. The system of equations is:
$$x + y = 12000$$
$$0.04x + 0.05y = 560$$
Multiplying the first equation by –0.04:
$$-0.04x - 0.04y = -480$$
$$0.04x + 0.05y = 560$$
Adding the two equations:
$$0.01y = 80$$
$$y = 8000$$
Substituting into the first equation:
$$x + 8000 = 12000$$
$$x = 4000$$
So $4,000 was invested at 4% and $8,000 was invested at 5%.

27. Let x represent the number of dimes and y represent the number of nickels. The system of equations is:
$$x + y = 17$$
$$0.10x + 0.05y = 1.35$$
Multiplying the first equation by –0.05:
$$-0.05x - 0.05y = -0.85$$
$$0.10x + 0.05y = 1.35$$
Adding the two equations:
$$0.05x = 0.50$$
$$x = 10$$
Substituting into the first equation:
$$10 + y = 17$$
$$y = 7$$
Barbara has 10 dimes and 7 nickels.

29. Let x represent the liters of 20% alcohol solution and y represent the liters of 10% alcohol solution. The system of equations is:
$$x + y = 50$$
$$0.20x + 0.10y = 0.12(50)$$
Multiplying the first equation by –0.10:
$$-0.10x - 0.10y = -5$$
$$0.20x + 0.10y = 6$$
Adding the two equations:
$$0.10x = 1$$
$$x = 10$$
Substituting into the first equation:
$$10 + y = 50$$
$$y = 40$$
The solution contains 40 liters of 10% alcohol solution and 10 liters of 20% alcohol solution.

Chapters 1-4 Cumulative Review

1. Simplifying using order of operations: $3 \cdot 4 + 5 = 12 + 5 = 17$

3. Simplifying using order of operations: $7[8 + (-5)] + 3(-7 + 12) = 7(3) + 3(5) = 21 + 15 = 36$

5. Simplifying using order of operations: $8 - 6(5 - 9) = 8 - 6(-4) = 8 + 24 = 32$

7. Simplifying: $\frac{2}{3} + \frac{3}{4} - \frac{1}{6} = \frac{2 \cdot 4}{3 \cdot 4} + \frac{3 \cdot 3}{4 \cdot 3} - \frac{1 \cdot 2}{6 \cdot 2} = \frac{8}{12} + \frac{9}{12} - \frac{2}{12} = \frac{15}{12} = \frac{5}{4}$

9. Solving the equation:

$$-5 - 6 = -y - 3 + 2y$$
$$-11 = y - 3$$
$$-11 + 3 = y - 3 + 3$$
$$y = -8$$

11. Solving the equation:
$$3(x - 4) = 9$$
$$3x - 12 = 9$$
$$3x - 12 + 12 = 9 + 12$$
$$3x = 21$$
$$\tfrac{1}{3}(3x) = \tfrac{1}{3}(21)$$
$$x = 7$$

13. Solving the inequality:
$$0.3x + 0.7 \le -2$$
$$0.3x + 0.7 - 0.7 \le -2 - 0.7$$
$$0.3x \le -2.7$$
$$\frac{0.3x}{0.3} \le \frac{-2.7}{0.3}$$
$$x \le -9$$

Graphing the solution set:

15. Graphing the line:

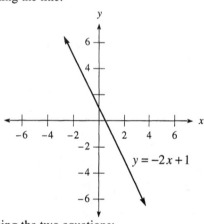

$y = -2x + 1$

17. Graphing the line:

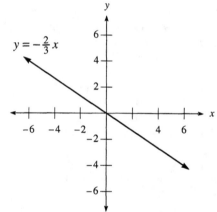

$y = -\frac{2}{3}x$

19. Graphing the two equations:

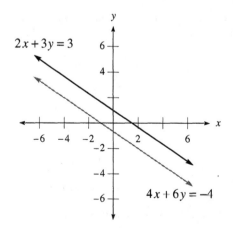

$2x + 3y = 3$

$4x + 6y = -4$

The two lines are parallel.

21. Multiplying the first equation by –2:
$$-2x - 2y = -14$$
$$2x + 2y = 14$$
Adding the two equations:
$$0 = 0$$
Since this statement is true, the system is dependent. The two lines coincide.

23. Multiplying the second equation by 3:
$$2x + 3y = 13$$
$$3x - 3y = -3$$
Adding the two equations:
$$5x = 10$$
$$x = 2$$
Substituting into the first equation:
$$2(2) + 3y = 13$$
$$4 + 3y = 13$$
$$3y = 9$$
$$y = 3$$
The solution is $(2,3)$.

25. Multiplying the second equation by –2:
$$2x + 5y = 33$$
$$-2x + 6y = 0$$
Adding the two equations:
$$11y = 33$$
$$y = 3$$
Substituting into the second equation:
$$x - 3(3) = 0$$
$$x - 9 = 0$$
$$x = 9$$
The solution is $(9,3)$.

27. Multiplying the second equation by 7:
$$3x - 7y = 12$$
$$14x + 7y = 56$$
Adding the two equations:
$$17x = 68$$
$$x = 4$$
Substituting into the second equation:
$$2(4) + y = 8$$
$$8 + y = 8$$
$$y = 0$$
The solution is $(4,0)$.

29. Substituting into the first equation:
$$2x - 3(5x + 2) = 7$$
$$2x - 15x - 6 = 7$$
$$-13x - 6 = 7$$
$$-13x = 13$$
$$x = -1$$
Substituting into the second equation: $y = 5(-1) + 2 = -5 + 2 = -3$. The solution is $(-1,-3)$.

31. commutative property of addition

33. The quotient is: $\dfrac{-30}{6} = -5$

35. Finding the percent:
$$p \bullet 82 = 20.5$$
$$p = \frac{20.5}{82}$$
$$p = 0.25 = 25\%$$
So 25% of 82 is 20.5.

37. Simplifying, then evaluating when $x = 3$: $-3x + 7 + 5x = 2x + 7 = 2(3) + 7 = 6 + 7 = 13$

39. Evaluating when $x = -2$: $4x - 5 = 4(-2) - 5 = -8 - 5 = -13$

41. The expression is: $5 - (-8) = 5 + 8 = 13$

43. To find the x-intercept, let $y = 0$:
$$3x - 4(0) = 12$$
$$3x = 12$$
$$x = 4$$
To find the y-intercept, let $x = 0$:
$$3(0) - 4y = 12$$
$$-4y = 12$$
$$y = -3$$

45. Computing the slope: $m = \dfrac{-4 - 1}{-5 - (-1)} = \dfrac{-5}{-4} = \frac{5}{4}$

47. The slope-intercept form is $y = \frac{2}{3}x + 3$.

49. First compute the slope: $m = \dfrac{6-3}{6-4} = \dfrac{3}{2}$. Using the point-slope formula:

$$y - 3 = \tfrac{3}{2}(x - 4)$$
$$y - 3 = \tfrac{3}{2}x - 6$$
$$y = \tfrac{3}{2}x - 3$$

Chapter 4 Test

1. Graphing the two equations:

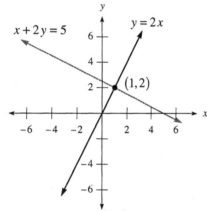

The intersection point is $(1, 2)$.

2. Graphing the two equations:

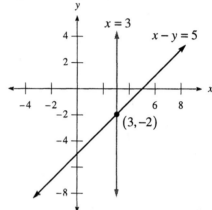

The intersection point is $(3, -2)$.

3. Adding the two equations:
$$3x = -9$$
$$x = -3$$
Substituting into the first equation:
$$-3 - y = 1$$
$$-y = 4$$
$$y = -4$$
The solution is $(-3, -4)$.

4. Multiplying the first equation by -1:
$$-2x - y = -7$$
$$3x + y = 12$$
Adding the two equations:
$$x = 5$$
Substituting into the first equation:
$$2(5) + y = 7$$
$$10 + y = 7$$
$$y = -3$$
The solution is $(5, -3)$.

5. Multiplying the second equation by 4:
$$7x + 8y = -2$$
$$12x - 8y = 40$$
Adding the two equations:
$$19x = 38$$
$$x = 2$$
Substituting into the first equation:
$$7(2) + 8y = -2$$
$$14 + 8y = -2$$
$$8y = -16$$
$$y = -2$$
The solution is $(2, -2)$.

6. Multiplying the first equation by -3 and the second equation by 2:
$$-18x + 30y = -18$$
$$18x - 30y = 18$$
Adding the two equations: $0 = 0$. Since this equation is true, the system is dependent. The two lines coincide.

7. Substituting into the first equation:
$$3x + 2(2x + 3) = 20$$
$$3x + 4x + 6 = 20$$
$$7x + 6 = 20$$
$$7x = 14$$
$$x = 2$$
Substituting into the second equation: $y = 2(2) + 3 = 4 + 3 = 7$. The solution is $(2, 7)$.

8. Substituting into the first equation:
$$3(y + 1) - 6y = -6$$
$$3y + 3 - 6y = -6$$
$$-3y + 3 = -6$$
$$-3y = -9$$
$$y = 3$$
Substituting into the second equation: $x = 3 + 1 = 4$. The solution is $(4, 3)$.

9. Solving the second equation for y:
$$-3x + y = 3$$
$$y = 3x + 3$$
Substituting into the first equation:
$$7x - 2(3x + 3) = -4$$
$$7x - 6x - 6 = -4$$
$$x - 6 = -4$$
$$x = 2$$
Substituting into $y = 3x + 3$: $y = 3(2) + 3 = 6 + 3 = 9$. The solution is $(2, 9)$.

10. Solving the second equation for x:
$$x + 3y = -8$$
$$x = -3y - 8$$
Substituting into the first equation:
$$2(-3y - 8) - 3y = -7$$
$$-6y - 16 - 3y = -7$$
$$-9y - 16 = -7$$
$$-9y = 9$$
$$y = -1$$
Substituting into $x = -3y - 8$: $x = -3(-1) - 8 = 3 - 8 = -5$. The solution is $(-5, -1)$.

11. Let x and y represent the two numbers. The system of equations is:
$$x + y = 12$$
$$x - y = 2$$
Adding the two equations:
$$2x = 14$$
$$x = 7$$
Substituting into the first equation:
$$7 + y = 12$$
$$y = 5$$
The two numbers are 5 and 7.

12. Let x and y represent the two numbers. The system of equations is:
$$x + y = 15$$
$$y = 6 + 2x$$
Substituting into the first equation:
$$x + 6 + 2x = 15$$
$$3x + 6 = 15$$
$$3x = 9$$
$$x = 3$$
Substituting into the second equation: $y = 6 + 2(3) = 6 + 6 = 12$. The two numbers are 3 and 12.

13.   Let x represent the amount invested at 9% and y represent the amount invested at 11%. The system of equations is:
$$x + y = 10000$$
$$0.09x + 0.11y = 980$$
Multiplying the first equation by –0.09:
$$-0.09x - 0.09y = -900$$
$$0.09x + 0.11y = 980$$
Adding the two equations:
$$0.02y = 80$$
$$y = 4000$$
Substituting into the first equation:
$$x + 4000 = 10000$$
$$x = 6000$$
Dr. Stork should invest $6,000 at 9%.

14.   Let x represent the number of nickels and y represent the number of quarters. The system of equations is:
$$x + y = 12$$
$$0.05x + 0.25y = 1.60$$
Multiplying the first equation by –0.05:
$$-0.05x - 0.05y = -0.60$$
$$0.05x + 0.25y = 1.60$$
Adding the two equations:
$$0.20y = 1.00$$
$$y = 5$$
Substituting into the first equation:
$$x + 5 = 12$$
$$x = 7$$
Diane has 7 nickels and 5 quarters.

Chapter 5
Exponents and Polynomials

5.1 Multiplication with Exponents

1. The base is 4 and the exponent is 2. Evaluating the expression: $4^2 = 4 \cdot 4 = 16$
3. The base is 0.3 and the exponent is 2. Evaluating the expression: $0.3^2 = 0.3 \cdot 0.3 = 0.09$
5. The base is 4 and the exponent is 3. Evaluating the expression: $4^3 = 4 \cdot 4 \cdot 4 = 64$
7. The base is –5 and the exponent is 2. Evaluating the expression: $(-5)^2 = (-5) \cdot (-5) = 25$
9. The base is 2 and the exponent is 3. Evaluating the expression: $-2^3 = -2 \cdot 2 \cdot 2 = -8$
11. The base is 3 and the exponent is 4. Evaluating the expression: $3^4 = 3 \cdot 3 \cdot 3 \cdot 3 = 81$
13. The base is $\frac{2}{3}$ and the exponent is 2. Evaluating the expression: $\left(\frac{2}{3}\right)^2 = \left(\frac{2}{3}\right) \cdot \left(\frac{2}{3}\right) = \frac{4}{9}$
15. The base is $\frac{1}{2}$ and the exponent is 4. Evaluating the expression: $\left(\frac{1}{2}\right)^4 = \left(\frac{1}{2}\right) \cdot \left(\frac{1}{2}\right) \cdot \left(\frac{1}{2}\right) \cdot \left(\frac{1}{2}\right) = \frac{1}{16}$
17. a. Completing the table:

Number (x)	1	2	3	4	5	6	7
Square $\left(x^2\right)$	1	4	9	16	25	36	49

  b. For numbers larger than 1, the square of the number is larger than the number.

19. Simplifying the expression: $x^4 \cdot x^5 = x^{4+5} = x^9$ 21. Simplifying the expression: $y^{10} \cdot y^{20} = y^{10+20} = y^{30}$
23. Simplifying the expression: $2^5 \cdot 2^4 \cdot 2^3 = 2^{5+4+3} = 2^{12}$
25. Simplifying the expression: $x^4 \cdot x^6 \cdot x^8 \cdot x^{10} = x^{4+6+8+10} = x^{28}$
27. Simplifying the expression: $\left(x^2\right)^5 = x^{2 \cdot 5} = x^{10}$ 29. Simplifying the expression: $\left(5^4\right)^3 = 5^{4 \cdot 3} = 5^{12}$
31. Simplifying the expression: $\left(y^3\right)^3 = y^{3 \cdot 3} = y^9$ 33. Simplifying the expression: $\left(2^5\right)^{10} = 2^{5 \cdot 10} = 2^{50}$
35. Simplifying the expression: $\left(a^3\right)^x = a^{3x}$ 37. Simplifying the expression: $\left(b^x\right)^y = b^{xy}$
39. Simplifying the expression: $(4x)^2 = 4^2 \cdot x^2 = 16x^2$ 41. Simplifying the expression: $(2y)^5 = 2^5 \cdot y^5 = 32y^5$
43. Simplifying the expression: $(-3x)^4 = (-3)^4 \cdot x^4 = 81x^4$
45. Simplifying the expression: $(0.5ab)^2 = (0.5)^2 \cdot a^2 b^2 = 0.25a^2 b^2$
47. Simplifying the expression: $(4xyz)^3 = 4^3 \cdot x^3 y^3 z^3 = 64x^3 y^3 z^3$
49. Simplifying using properties of exponents: $\left(2x^4\right)^3 = 2^3 \left(x^4\right)^3 = 8x^{12}$
51. Simplifying using properties of exponents: $\left(4a^3\right)^2 = 4^2 \left(a^3\right)^2 = 16a^6$
53. Simplifying using properties of exponents: $\left(x^2\right)^3 \left(x^4\right)^2 = x^6 \cdot x^8 = x^{14}$
55. Simplifying using properties of exponents: $\left(a^3\right)^1 \left(a^2\right)^4 = a^3 \cdot a^8 = a^{11}$

57. Simplifying using properties of exponents: $(2x)^3 (2x)^4 = (2x)^7 = 2^7 x^7 = 128x^7$

59. Simplifying using properties of exponents: $(3x^2)^3 (2x)^4 = 3^3 x^6 \cdot 2^4 x^4 = 27x^6 \cdot 16x^4 = 432x^{10}$

61. Simplifying using properties of exponents: $(4x^2 y^3)^2 = 4^2 x^4 y^6 = 16x^4 y^6$

63. Simplifying using properties of exponents: $\left(\frac{2}{3} a^4 b^5\right)^3 = \left(\frac{2}{3}\right)^3 a^{12} b^{15} = \frac{8}{27} a^{12} b^{15}$

65. **a.** Completing the table:

Number (x)	Square (x^2)
−3	9
−2	4
−1	1
0	0
1	1
2	4
3	9

b. Constructing a line graph:

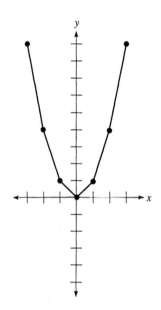

67. Completing the table:

Number (x)	Square (x^2)
−2.5	6.25
−1.5	2.25
−0.5	0.25
0	0
0.5	0.25
1.5	2.25
2.5	6.25

69. Writing in scientific notation: $43,200 = 4.32 \times 10^4$

71. Writing in scientific notation: $570 = 5.7 \times 10^2$

73. Writing in scientific notation: $238,000 = 2.38 \times 10^5$

75. Writing in expanded form: $2.49 \times 10^3 = 2,490$

77. Writing in expanded form: $3.52 \times 10^2 = 352$

79. Writing in expanded form: $2.8 \times 10^4 = 28,000$

81. The volume is given by: $V = (3 \text{ in.})^3 = 27 \text{ inches}^3$

83. The volume is given by: $V = (2.5 \text{ in.})^3 \approx 15.6 \text{ inches}^3$

85. The volume is given by: $V = (8 \text{ in.})(4.5 \text{ in.})(1 \text{ in.}) = 36 \text{ inches}^3$

87. Yes. If the box was 7 ft x 3 ft x 2 ft, you could fit inside of it.

89. Writing in scientific notation: $650,000,000 \text{ seconds} = 6.5 \times 10^8$ seconds

91. Writing in expanded form: 7.4×10^5 dollars $= \$740,000$

93. Writing in expanded form: 1.8×10^5 dollars $= \$180,000$

95. Substitute $c = 8$, $b = 3.35$, and $s = 3.11$: $d = \pi \cdot 3.11 \cdot 8 \cdot \left(\frac{1}{2} \cdot 3.35\right)^2 \approx 219 \text{ inches}^3$

97. Substitute $c = 6$, $b = 3.59$, and $s = 2.99$: $d = \pi \cdot 2.99 \cdot 6 \cdot \left(\frac{1}{2} \cdot 3.59\right)^2 \approx 182 \text{ inches}^3$

99. **a.** Each section is $\frac{1}{3}$ foot, so after stage 1 the length is $\frac{4}{3}$ feet.

    b. After stage 2 the length is $\left(\frac{4}{3}\right)^2 = \frac{16}{9}$ feet.   **c.** After stage 3 the length is $\left(\frac{4}{3}\right)^3 = \frac{64}{27}$ feet.

    d. After stage 10 the length is $\left(\frac{4}{3}\right)^{10}$ feet.

101. Subtracting: $4 - 7 = 4 + (-7) = -3$ **103.** Subtracting: $4 - (-7) = 4 + 7 = 11$

105. Subtracting: $15 - 20 = 15 + (-20) = -5$ **107.** Subtracting: $-15 - (-20) = -15 + 20 = 5$

5.2 Division with Exponents

1. Writing with positive exponents: $3^{-2} = \frac{1}{3^2} = \frac{1}{9}$ **3.** Writing with positive exponents: $6^{-2} = \frac{1}{6^2} = \frac{1}{36}$

5. Writing with positive exponents: $8^{-2} = \frac{1}{8^2} = \frac{1}{64}$ **7.** Writing with positive exponents: $5^{-3} = \frac{1}{5^3} = \frac{1}{125}$

9. Writing with positive exponents: $2x^{-3} = 2 \cdot \frac{1}{x^3} = \frac{2}{x^3}$

11. Writing with positive exponents: $(2x)^{-3} = \frac{1}{(2x)^3} = \frac{1}{8x^3}$

13. Writing with positive exponents: $(5y)^{-2} = \frac{1}{(5y)^2} = \frac{1}{25y^2}$

15. Writing with positive exponents: $10^{-2} = \frac{1}{10^2} = \frac{1}{100}$

17. Completing the table:

Number (x)	Square $\left(x^2\right)$	Power of 2 $\left(2^x\right)$
−3	9	$\frac{1}{8}$
−2	4	$\frac{1}{4}$
−1	1	$\frac{1}{2}$
0	0	1
1	1	2
2	4	4
3	9	8

19. Simplifying: $\frac{5^1}{5^3} = 5^{1-3} = 5^{-2} = \frac{1}{5^2} = \frac{1}{25}$ **21.** Simplifying: $\frac{x^{10}}{x^4} = x^{10-4} = x^6$

23. Simplifying: $\frac{4^3}{4^0} = 4^{3-0} = 4^3 = 64$ **25.** Simplifying: $\frac{(2x)^7}{(2x)^4} = (2x)^{7-4} = (2x)^3 = 2^3 x^3 = 8x^3$

27. Simplifying: $\frac{6^{11}}{6} = \frac{6^{11}}{6^1} = 6^{11-1} = 6^{10}$ $(= 60,466,176)$

29. Simplifying: $\frac{6}{6^{11}} = \frac{6^1}{6^{11}} = 6^{1-11} = 6^{-10} = \frac{1}{6^{10}}$ $\left(= \frac{1}{60,466,176}\right)$

31.   Simplifying: $\dfrac{2^{-5}}{2^3} = 2^{-5-3} = 2^{-8} = \dfrac{1}{2^8} = \dfrac{1}{256}$

33.   Simplifying: $\dfrac{2^5}{2^{-3}} = 2^{5-(-3)} = 2^{5+3} = 2^8 = 256$

35.   Simplifying: $\dfrac{(3x)^{-5}}{(3x)^{-8}} = (3x)^{-5-(-8)} = (3x)^{-5+8} = (3x)^3 = 3^3 x^3 = 27x^3$

37.   Simplifying: $(3xy)^4 = 3^4 x^4 y^4 = 81x^4 y^4$

39.   Simplifying: $10^0 = 1$

41.   Simplifying: $\left(2a^2 b\right)^1 = 2a^2 b$

43.   Simplifying: $\left(7y^3\right)^{-2} = \dfrac{1}{\left(7y^3\right)^2} = \dfrac{1}{49y^6}$

45.   Simplifying: $x^{-3} \bullet x^{-5} = x^{-3-5} = x^{-8} = \dfrac{1}{x^8}$

47.   Simplifying: $y^7 \bullet y^{-10} = y^{7-10} = y^{-3} = \dfrac{1}{y^3}$

49.   Simplifying: $\dfrac{\left(x^2\right)^3}{x^4} = \dfrac{x^6}{x^4} = x^{6-4} = x^2$

51.   Simplifying: $\dfrac{\left(a^4\right)^3}{\left(a^3\right)^2} = \dfrac{a^{12}}{a^6} = a^{12-6} = a^6$

53.   Simplifying: $\dfrac{y^7}{\left(y^2\right)^8} = \dfrac{y^7}{y^{16}} = y^{7-16} = y^{-9} = \dfrac{1}{y^9}$

55.   Simplifying: $\left(\dfrac{y^7}{y^2}\right)^8 = \left(y^{7-2}\right)^8 = \left(y^5\right)^8 = y^{40}$

57.   Simplifying: $\dfrac{\left(x^{-2}\right)^3}{x^{-5}} = \dfrac{x^{-6}}{x^{-5}} = x^{-6-(-5)} = x^{-6+5} = x^{-1} = \dfrac{1}{x}$

59.   Simplifying: $\left(\dfrac{x^{-2}}{x^{-5}}\right)^3 = \left(x^{-2+5}\right)^3 = \left(x^3\right)^3 = x^9$

61.   Simplifying: $\dfrac{\left(a^3\right)^2 \left(a^4\right)^5}{\left(a^5\right)^2} = \dfrac{a^6 \bullet a^{20}}{a^{10}} = \dfrac{a^{26}}{a^{10}} = a^{26-10} = a^{16}$

63.   Simplifying: $\dfrac{\left(a^{-2}\right)^3 \left(a^4\right)^2}{\left(a^{-3}\right)^{-2}} = \dfrac{a^{-6} \bullet a^8}{a^6} = \dfrac{a^2}{a^6} = a^{2-6} = a^{-4} = \dfrac{1}{a^4}$

65.   Completing the table:

Number (x)	Power of 2 $\left(2^x\right)$
-3	$\frac{1}{8}$
-2	$\frac{1}{4}$
-1	$\frac{1}{2}$
0	1
1	2
2	4
3	8

Constructing the line graph:

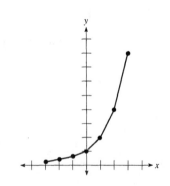

67. Writing in scientific notation: $0.0048 = 4.8 \times 10^{-3}$ **69.** Writing in scientific notation: $25 = 2.5 \times 10^1$

71. Writing in scientific notation: $0.000009 = 9 \times 10^{-6}$

73. Completing the table:

Expanded Form	Scientific Notation $\left(n \times 10^r\right)$
0.000357	3.57×10^{-4}
0.00357	3.57×10^{-3}
0.0357	3.57×10^{-2}
0.357	3.57×10^{-1}
3.57	3.57×10^0
35.7	3.57×10^1
357	3.57×10^2
3,570	3.57×10^3
35,700	3.57×10^4

75. Writing in expanded form: $4.23 \times 10^{-3} = 0.00423$ **77.** Writing in expanded form: $8 \times 10^{-5} = 0.00008$

79. Writing in expanded form: $4.2 \times 10^0 = 4.2$

81. Writing in expanded form: 2×10^{-3} seconds $= 0.002$ seconds

83. Writing in scientific notation: 0.006 inches $= 6 \times 10^{-3}$ inches

85. Writing in scientific notation: $25 \times 10^3 = 2.5 \times 10^4$ **87.** Writing in scientific notation: $23.5 \times 10^4 = 2.35 \times 10^5$

89. Writing in scientific notation: $0.82 \times 10^{-3} = 8.2 \times 10^{-4}$

91. The area of the smaller square is $(10 \text{ in.})^2 = 100$ inches^2, while the area of the larger square is $(20 \text{ in.})^2 = 400$ inches^2. It would take 4 smaller squares to cover the larger square.

93. The area of the smaller square is x^2, while the area of the larger square is $(2x)^2 = 4x^2$. It would take 4 smaller squares to cover the larger square.

95. The volume of the smaller box is $(6 \text{ in.})^3 = 216$ inches^3, while the volume of the larger box is $(12 \text{ in.})^3 = 1,728$ inches^3. Thus 8 smaller boxes will fit inside the larger box ($8 \cdot 216 = 1,728$).

97. The volume of the smaller box is x^3, while the volume of the larger box is $(2x)^3 = 8x^3$. Thus 8 smaller boxes will fit inside the larger box.

99. Simplifying by combining like terms: $4x + 3x = (4+3)x = 7x$

101. Simplifying by combining like terms: $5a - 3a = (5-3)a = 2a$

103. Simplifying by combining like terms: $4y + 5y + y = (4+5+1)y = 10y$

5.3 Operations with Monomials

1. Multiplying the monomials: $\left(3x^4\right)\left(4x^3\right) = 12x^{4+3} = 12x^7$

3. Multiplying the monomials: $\left(-2y^4\right)\left(8y^7\right) = -16y^{4+7} = -16y^{11}$

5. Multiplying the monomials: $(8x)(4x) = 32x^{1+1} = 32x^2$

7. Multiplying the monomials: $\left(10a^3\right)(10a)\left(2a^2\right) = 200a^{3+1+2} = 200a^6$

9. Multiplying the monomials: $\left(6ab^2\right)\left(-4a^2b\right) = -24a^{1+2}b^{2+1} = -24a^3b^3$

11. Multiplying the monomials: $\left(4x^2y\right)\left(3x^3y^3\right)\left(2xy^4\right) = 24x^{2+3+1}y^{1+3+4} = 24x^6y^8$

13. Dividing the monomials: $\dfrac{15x^3}{5x^2} = \dfrac{15}{5} \cdot \dfrac{x^3}{x^2} = 3x$

15. Dividing the monomials: $\dfrac{18y^9}{3y^{12}} = \dfrac{18}{3} \cdot \dfrac{y^9}{y^{12}} = 6 \cdot \dfrac{1}{y^3} = \dfrac{6}{y^3}$

17.   Dividing the monomials: $\dfrac{32a^3}{64a^4} = \dfrac{32}{64} \cdot \dfrac{a^3}{a^4} = \dfrac{1}{2} \cdot \dfrac{1}{a} = \dfrac{1}{2a}$

19.   Dividing the monomials: $\dfrac{21a^2b^3}{-7ab^5} = \dfrac{21}{-7} \cdot \dfrac{a^2}{a} \cdot \dfrac{b^3}{b^5} = -3 \cdot a \cdot \dfrac{1}{b^2} = -\dfrac{3a}{b^2}$

21.   Dividing the monomials: $\dfrac{3x^3y^2z}{27xy^2z^3} = \dfrac{3}{27} \cdot \dfrac{x^3}{x} \cdot \dfrac{y^2}{y^2} \cdot \dfrac{z}{z^3} = \dfrac{1}{9} \cdot x^2 \cdot \dfrac{1}{z^2} = \dfrac{x^2}{9z^2}$

23.   Completing the table:

a	b	ab	$\dfrac{a}{b}$	$\dfrac{b}{a}$
10	$5x$	$50x$	$\dfrac{2}{x}$	$\dfrac{x}{2}$
$20x^3$	$6x^2$	$120x^5$	$\dfrac{10x}{3}$	$\dfrac{3}{10x}$
$25x^5$	$5x^4$	$125x^9$	$5x$	$\dfrac{1}{5x}$
$3x^{-2}$	$3x^2$	9	$\dfrac{1}{x^4}$	x^4
$-2y^4$	$8y^7$	$-16y^{11}$	$-\dfrac{1}{4y^3}$	$-4y^3$

25.   Finding the product: $\left(3\times10^3\right)\left(2\times10^5\right) = 6\times10^8$

27.   Finding the product: $\left(3.5\times10^4\right)\left(5\times10^{-6}\right) = 17.5\times10^{-2} = 1.75\times10^{-1}$

29.   Finding the product: $\left(5.5\times10^{-3}\right)\left(2.2\times10^{-4}\right) = 12.1\times10^{-7} = 1.21\times10^{-6}$

31.   Finding the quotient: $\dfrac{8.4\times10^5}{2\times10^2} = 4.2\times10^3$

33.   Finding the quotient: $\dfrac{6\times10^8}{2\times10^{-2}} = 3\times10^{10}$

35.   Finding the quotient: $\dfrac{2.5\times10^{-6}}{5\times10^{-4}} = 0.5\times10^{-2} = 5.0\times10^{-3}$

37.   Combining the monomials: $3x^2 + 5x^2 = (3+5)x^2 = 8x^2$

39.   Combining the monomials: $8x^5 - 19x^5 = (8-19)x^5 = -11x^5$

41.   Combining the monomials: $2a + a - 3a = (2+1-3)a = 0a = 0$

43.   Combining the monomials: $10x^3 - 8x^3 + 2x^3 = (10-8+2)x^3 = 4x^3$

45.   Combining the monomials: $20ab^2 - 19ab^2 + 30ab^2 = (20-19+30)ab^2 = 31ab^2$

47.   Completing the table:

a	b	ab	$a+b$
$5x$	$3x$	$15x^2$	$8x$
$4x^2$	$2x^2$	$8x^4$	$6x^2$
$3x^3$	$6x^3$	$18x^6$	$9x^3$
$2x^4$	$-3x^4$	$-6x^8$	$-x^4$
x^5	$7x^5$	$7x^{10}$	$8x^5$

49.   Simplifying the expression: $\dfrac{\left(3x^2\right)\left(8x^5\right)}{6x^4} = \dfrac{24x^7}{6x^4} = \dfrac{24}{6} \cdot \dfrac{x^7}{x^4} = 4x^3$

51.   Simplifying the expression: $\dfrac{\left(9a^2b\right)\left(2a^3b^4\right)}{18a^5b^7} = \dfrac{18a^5b^5}{18a^5b^7} = \dfrac{18}{18} \cdot \dfrac{a^5}{a^5} \cdot \dfrac{b^5}{b^7} = 1 \cdot \dfrac{1}{b^2} = \dfrac{1}{b^2}$

53.   Simplifying the expression: $\dfrac{\left(4x^3y^2\right)\left(9x^4y^{10}\right)}{\left(3x^5y\right)\left(2x^6y\right)} = \dfrac{36x^7y^{12}}{6x^{11}y^2} = \dfrac{36}{6} \cdot \dfrac{x^7}{x^{11}} \cdot \dfrac{y^{12}}{y^2} = 6 \cdot \dfrac{1}{x^4} \cdot y^{10} = \dfrac{6y^{10}}{x^4}$

55. Simplifying the expression: $\dfrac{\left(6\times10^8\right)\left(3\times10^5\right)}{9\times10^7}=\dfrac{18\times10^{13}}{9\times10^7}=2\times10^6$

57. Simplifying the expression: $\dfrac{\left(5\times10^3\right)\left(4\times10^{-5}\right)}{2\times10^{-2}}=\dfrac{20\times10^{-2}}{2\times10^{-2}}=10=1\times10^1$

59. Simplifying the expression: $\dfrac{\left(2.8\times10^{-7}\right)\left(3.6\times10^4\right)}{2.4\times10^3}=\dfrac{10.08\times10^{-3}}{2.4\times10^3}=4.2\times10^{-6}$

61. Simplifying the expression: $\dfrac{18x^4}{3x}+\dfrac{21x^7}{7x^4}=6x^3+3x^3=9x^3$

63. Simplifying the expression: $\dfrac{45a^6}{9a^4}-\dfrac{50a^8}{2a^6}=5a^2-25a^2=-20a^2$

65. Simplifying the expression: $\dfrac{6x^7y^4}{3x^2y^2}+\dfrac{8x^5y^8}{2y^6}=2x^5y^2+4x^5y^2=6x^5y^2$

67. Solving the equation:

$$4^x\bullet4^5=4^7$$
$$4^{x+5}=4^7$$
$$x+5=7$$
$$x=2$$

69. Solving the equation:

$$\left(7^3\right)^x=7^{12}$$
$$7^{3x}=7^{12}$$
$$3x=12$$
$$x=4$$

71. Simplifying each value:

$$(a+b)^2=(4+5)^2=9^2=81$$
$$a^2+b^2=4^2+5^2=16+25=41$$

Note that the values are not equal.

73. Simplifying each value:

$$(a+b)^2=(3+4)^2=7^2=49 \qquad a^2+2ab+b^2=3^2+2(3)(4)+4^2=9+24+16=49$$

Note that the values are equal.

75. Let x represent the width and $2x$ represent the length. The perimeter and area are given by:

$$P=2(x)+2(2x)=2x+4x=6x \qquad A=x\bullet2x=2x^2$$

77. Let x represent the width and $2x$ represent the length. The volume is given by: $V=(x)(2x)(4)=8x^2$ inches^3

79. **a.** The volume is given by: $V=(8.5)(55)(10.1)\approx4,700$ feet$^3=4.7\times10^3$ feet^3

  b. Dividing: $\dfrac{4.7\times10^3}{0.15}\approx31,478$ boxes

81. Evaluating when $x=-2$: $4x=4(-2)=-8$

83. Evaluating when $x=-2$: $-2x+5=-2(-2)+5=4+5=9$

85. Evaluating when $x=-2$: $x^2+5x+6=(-2)^2+5(-2)+6=4-10+6=0$

87. The ordered pairs are $(-2,-2)$, $(0,2)$, and $(2,6)$:

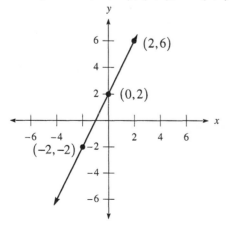

89. The ordered pairs are $(-3,0)$, $(0,1)$, and $(3,2)$:

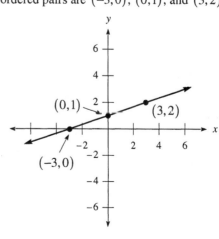

5.4 Addition and Subtraction of Polynomials

1. This is a trinomial of degree 3.　　　　　**3.** This is a trinomial of degree 3.
5. This is a binomial of degree 1.　　　　　**7.** This is a binomial of degree 2.
9. This is a monomial of degree 2.　　　　**11.** This is a monomial of degree 0.

13. Combining the polynomials: $(2x^2+3x+4)+(3x^2+2x+5)=(2x^2+3x^2)+(3x+2x)+(4+5)=5x^2+5x+9$

15. Combining the polynomials: $(3a^2-4a+1)+(2a^2-5a+6)=(3a^2+2a^2)+(-4a-5a)+(1+6)=5a^2-9a+7$

17. Combining the polynomials: $x^2+4x+2x+8=x^2+(4x+2x)+8=x^2+6x+8$

19. Combining the polynomials: $6x^2-3x-10x+5=6x^2+(-3x-10x)+5=6x^2-13x+5$

21. Combining the polynomials: $x^2-3x+3x-9=x^2+(-3x+3x)-9=x^2-9$

23. Combining the polynomials: $3y^2-5y-6y+10=3y^2+(-5y-6y)+10=3y^2-11y+10$

25. Combining the polynomials:
$$(6x^3-4x^2+2x)+(9x^2-6x+3)=6x^3+(-4x^2+9x^2)+(2x-6x)+3=6x^3+5x^2-4x+3$$

27. Combining the polynomials:
$$\left(\tfrac{2}{3}x^2-\tfrac{1}{5}x-\tfrac{3}{4}\right)+\left(\tfrac{4}{3}x^2-\tfrac{4}{5}x+\tfrac{7}{4}\right)=\left(\tfrac{2}{3}x^2+\tfrac{4}{3}x^2\right)+\left(-\tfrac{1}{5}x-\tfrac{4}{5}x\right)+\left(-\tfrac{3}{4}+\tfrac{7}{4}\right)=2x^2-x+1$$

29. Combining the polynomials: $(a^2-a-1)-(-a^2+a+1)=a^2-a-1+a^2-a-1=2a^2-2a-2$

31. Combining the polynomials:
$$\left(\tfrac{5}{9}x^3+\tfrac{1}{3}x^2-2x+1\right)-\left(\tfrac{2}{3}x^3+x^2+\tfrac{1}{2}x-\tfrac{3}{4}\right)=\tfrac{5}{9}x^3+\tfrac{1}{3}x^2-2x+1-\tfrac{2}{3}x^3-x^2-\tfrac{1}{2}x+\tfrac{3}{4}$$
$$=-\tfrac{1}{9}x^3-\tfrac{2}{3}x^2-\tfrac{5}{2}x+\tfrac{7}{4}$$

33. Combining the polynomials:
$$(4y^2-3y+2)+(5y^2+12y-4)-(13y^2-6y+20)=4y^2-3y+2+5y^2+12y-4-13y^2+6y-20$$
$$=(4y^2+5y^2-13y^2)+(-3y+12y+6y)+(2-4-20)$$
$$=-4y^2+15y-22$$

35. Performing the subtraction:
$$(11x^2-10x+13)-(10x^2+23x-50)=11x^2-10x+13-10x^2-23x+50$$
$$=(11x^2-10x^2)+(-10x-23x)+(13+50)$$
$$=x^2-33x+63$$

37. Performing the subtraction:
$$(11y^2+11y+11)-(3y^2+7y-15)=11y^2+11y+11-3y^2-7y+15$$
$$=(11y^2-3y^2)+(11y-7y)+(11+15)$$
$$=8y^2+4y+26$$

39. Performing the addition:
$$(25x^2-50x+75)+(50x^2-100x-150)=(25x^2+50x^2)+(-50x-100x)+(75-150)=75x^2-150x-75$$

41. Performing the operations:
$$(3x-2)+(11x+5)-(2x+1)=3x-2+11x+5-2x-1=(3x+11x-2x)+(-2+5-1)=12x+2$$

43. Evaluating when $x=3$: $x^2-2x+1=(3)^2-2(3)+1=9-6+1=4$

45. Evaluating when $y=10$: $(y-5)^2=(10-5)^2=(5)^2=25$

47. Evaluating when $a=2$: $a^2+4a+4=(2)^2+4(2)+4=4+8+4=16$

49. Finding the volume of the cylinder and sphere:
$$V_{\text{cylinder}}=\pi(3^2)(6)=54\pi \qquad V_{\text{sphere}}=\tfrac{4}{3}\pi(3^3)=36\pi$$
Subtracting to find the amount of space to pack: $V=54\pi-36\pi=18\pi$ inches^3

51. Multiplying the monomials: $3x(-5x) = -15x^2$

53. Multiplying the monomials: $2x\left(3x^2\right) = 6x^{1+2} = 6x^3$

55. Multiplying the monomials: $3x^2\left(2x^2\right) = 6x^{2+2} = 6x^4$

5.5 Multiplication with Polynomials

1. Using the distributive property: $2x(3x+1) = 2x(3x) + 2x(1) = 6x^2 + 2x$

3. Using the distributive property: $2x^2\left(3x^2 - 2x + 1\right) = 2x^2\left(3x^2\right) - 2x^2(2x) + 2x^2(1) = 6x^4 - 4x^3 + 2x^2$

5. Using the distributive property: $2ab\left(a^2 - ab + 1\right) = 2ab\left(a^2\right) - 2ab(ab) + 2ab(1) = 2a^3b - 2a^2b^2 + 2ab$

7. Using the distributive property: $y^2\left(3y^2 + 9y + 12\right) = y^2\left(3y^2\right) + y^2(9y) + y^2(12) = 3y^4 + 9y^3 + 12y^2$

9. Using the distributive property:
$$4x^2y\left(2x^3y + 3x^2y^2 + 8y^3\right) = 4x^2y\left(2x^3y\right) + 4x^2y\left(3x^2y^2\right) + 4x^2y\left(8y^3\right) = 8x^5y^2 + 12x^4y^3 + 32x^2y^4$$

11. Multiplying using the FOIL method: $(x+3)(x+4) = x^2 + 3x + 4x + 12 = x^2 + 7x + 12$

13. Multiplying using the FOIL method: $(x+6)(x+1) = x^2 + 6x + 1x + 6 = x^2 + 7x + 6$

15. Multiplying using the FOIL method: $\left(x+\frac{1}{2}\right)\left(x+\frac{3}{2}\right) = x^2 + \frac{1}{2}x + \frac{3}{2}x + \frac{3}{4} = x^2 + 2x + \frac{3}{4}$

17. Multiplying using the FOIL method: $(a+5)(a-3) = a^2 + 5a - 3a - 15 = a^2 + 2a - 15$

19. Multiplying using the FOIL method: $(x-a)(y+b) = xy - ay + bx - ab$

21. Multiplying using the FOIL method: $(x+6)(x-6) = x^2 + 6x - 6x - 36 = x^2 - 36$

23. Multiplying using the FOIL method: $\left(y+\frac{5}{6}\right)\left(y-\frac{5}{6}\right) = y^2 + \frac{5}{6}y - \frac{5}{6}y - \frac{25}{36} = y^2 - \frac{25}{36}$

25. Multiplying using the FOIL method: $(2x-3)(x-4) = 2x^2 - 3x - 8x + 12 = 2x^2 - 11x + 12$

27. Multiplying using the FOIL method: $(a+2)(2a-1) = 2a^2 + 4a - a - 2 = 2a^2 + 3a - 2$

29. Multiplying using the FOIL method: $(2x-5)(3x-2) = 6x^2 - 15x - 4x + 10 = 6x^2 - 19x + 10$

31. Multiplying using the FOIL method: $(2x+3)(a+4) = 2ax + 3a + 8x + 12$

33. Multiplying using the FOIL method: $(5x-4)(5x+4) = 25x^2 - 20x + 20x - 16 = 25x^2 - 16$

35. Multiplying using the FOIL method: $\left(2x-\frac{1}{2}\right)\left(x+\frac{3}{2}\right) = 2x^2 - \frac{1}{2}x + 3x - \frac{3}{4} = 2x^2 + \frac{5}{2}x - \frac{3}{4}$

37. Multiplying using the FOIL method: $(1-2a)(3-4a) = 3 - 6a - 4a + 8a^2 = 3 - 10a + 8a^2$

39. The product is $(x+2)(x+3) = x^2 + 5x + 6$:

	x	3
x	x^2	$3x$
2	$2x$	6

41. The product is $(x+1)(2x+2) = 2x^2 + 4x + 2$:

	x	x	2
x	x^2	x^2	$2x$
1	x	x	2

43. Multiplying using the column method:

$$
\begin{array}{r}
a^2 - 3a + 2 \\
a - 3 \\
\hline
a^3 - 3a^2 + 2a \\
-3a^2 + 9a - 6 \\
\hline
a^3 - 6a^2 + 11a - 6
\end{array}
$$

45. Multiplying using the column method:

$$
\begin{array}{r}
x^2 - 2x + 4 \\
x + 2 \\
\hline
x^3 - 2x^2 + 4x \\
2x^2 - 4x + 8 \\
\hline
x^3 + 8
\end{array}
$$

47. Multiplying using the column method:

$$
\begin{array}{r}
x^2 +8x +9 \\
2x +1 \\
\hline
2x^3 +16x^2 +18x \\
x^2 +8x +9 \\
\hline
2x^3 +17x^2 +26x +9
\end{array}
$$

49. Multiplying using the column method:

$$
\begin{array}{r}
5x^2 +2x +1 \\
x^2 -3x +5 \\
\hline
5x^4 +2x^3 + x^2 \\
-15x^3 -6x^2 -3x \\
25x^2 +10x +5 \\
\hline
5x^4 -13x^3 +20x^2 +7x +5
\end{array}
$$

51. Multiplying using the FOIL method: $\left(x^2 +3\right)\left(2x^2 -5\right) = 2x^4 -5x^2 +6x^2 -15 = 2x^4 +x^2 -15$

53. Multiplying using the FOIL method: $\left(3a^4 +2\right)\left(2a^2 +5\right) = 6a^6 +15a^4 +4a^2 +10$

55. First multiply two polynomials using the FOIL method: $(x+3)(x+4) = x^2 +3x +4x +12 = x^2 +7x +12$
Now using the column method:

$$
\begin{array}{r}
x^2 +7x +12 \\
x +5 \\
\hline
x^3 +7x^2 +12x \\
5x^2 +35x +60 \\
\hline
x^3 +12x^2 +47x +60
\end{array}
$$

57. Let x represent the width and $2x + 5$ represent the length. The area is given by: $A = x(2x+5) = 2x^2 +5x$

59. Let x and $x + 1$ represent the width and length, respectively. The area is given by: $A = x(x+1) = x^2 +x$

61. The revenue is: $R = xp = (1200 -100p)p = 1200p -100p^2$

63. The revenue is: $R = xp = (1700 -100p)p = 1700p -100p^2$

65. Graphing each line:

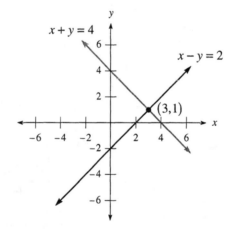

The intersection point is $(3,1)$.

67. Multiplying the first equation by 3 and the second equation by –2:
$$6x +9y = -3$$
$$-6x -10y = 4$$
Adding the two equations:
$$-y = 1$$
$$y = -1$$
Substituting into the first equation:
$$2x +3(-1) = -1$$
$$2x -3 = -1$$
$$2x = 2$$
$$x = 1$$
The solution is $(1,-1)$.

69.   Substituting into the first equation:
$$2x - 6(3x + 1) = 2$$
$$2x - 18x - 6 = 2$$
$$-16x - 6 = 2$$
$$-16x = 8$$
$$x = -\tfrac{1}{2}$$

Substituting into the second equation: $y = 3\left(-\tfrac{1}{2}\right) + 1 = -\tfrac{3}{2} + 1 = -\tfrac{1}{2}$. The solution is $\left(-\tfrac{1}{2}, -\tfrac{1}{2}\right)$.

71.   Let x represent the number of dimes and y represent the number of quarters. The system of equations is:
$$x + y = 11$$
$$0.10x + 0.25y = 1.85$$
Multiplying the first equation by –0.10:
$$-0.10x - 0.10y = -1.10$$
$$0.10x + 0.25y = 1.85$$
Adding the two equations:
$$0.15y = 0.75$$
$$y = 5$$
Substituting into the first equation:
$$x + 5 = 11$$
$$x = 6$$
Amy has 6 dimes and 5 quarters.

5.6 Binomial Squares and Other Special Products

1.   Multiplying using the FOIL method: $(x - 2)^2 = (x - 2)(x - 2) = x^2 - 2x - 2x + 4 = x^2 - 4x + 4$

3.   Multiplying using the FOIL method: $(a + 3)^2 = (a + 3)(a + 3) = a^2 + 3a + 3a + 9 = a^2 + 6a + 9$

5.   Multiplying using the FOIL method: $(x - 5)^2 = (x - 5)(x - 5) = x^2 - 5x - 5x + 25 = x^2 - 10x + 25$

7.   Multiplying using the FOIL method: $\left(a - \tfrac{1}{2}\right)^2 = \left(a - \tfrac{1}{2}\right)\left(a - \tfrac{1}{2}\right) = a^2 - \tfrac{1}{2}a - \tfrac{1}{2}a + \tfrac{1}{4} = a^2 - a + \tfrac{1}{4}$

9.   Multiplying using the FOIL method: $(x + 10)^2 = (x + 10)(x + 10) = x^2 + 10x + 10x + 100 = x^2 + 20x + 100$

11.   Multiplying using the square of binomial formula: $(a + 0.8)^2 = a^2 + 2(a)(0.8) + (0.8)^2 = a^2 + 1.6a + 0.64$

13.   Multiplying using the square of binomial formula: $(2x - 1)^2 = (2x)^2 - 2(2x)(1) + (1)^2 = 4x^2 - 4x + 1$

15.   Multiplying using the square of binomial formula: $(4a + 5)^2 = (4a)^2 + 2(4a)(5) + (5)^2 = 16a^2 + 40a + 25$

17.   Multiplying using the square of binomial formula: $(3x - 2)^2 = (3x)^2 - 2(3x)(2) + (2)^2 = 9x^2 - 12x + 4$

19.   Multiplying using the square of binomial formula: $(3a + 5b)^2 = (3a)^2 + 2(3a)(5b) + (5b)^2 = 9a^2 + 30ab + 25b^2$

21.   Multiplying using the square of binomial formula: $(4x - 5y)^2 = (4x)^2 - 2(4x)(5y) + (5y)^2 = 16x^2 - 40xy + 25y^2$

23.   Multiplying using the square of binomial formula: $(7m + 2n)^2 = (7m)^2 + 2(7m)(2n) + (2n)^2 = 49m^2 + 28mn + 4n^2$

25.   Multiplying using the square of binomial formula:
$$(6x - 10y)^2 = (6x)^2 - 2(6x)(10y) + (10y)^2 = 36x^2 - 120xy + 100y^2$$

27.   Multiplying using the square of binomial formula: $\left(x^2 + 5\right)^2 = \left(x^2\right)^2 + 2\left(x^2\right)(5) + (5)^2 = x^4 + 10x^2 + 25$

29.   Multiplying using the square of binomial formula: $\left(a^2 + 1\right)^2 = \left(a^2\right)^2 + 2\left(a^2\right)(1) + (1)^2 = a^4 + 2a^2 + 1$

31.   Completing the table:

x	$(x+3)^2$	$x^2 + 9$	$x^2 + 6x + 9$
1	16	10	16
2	25	13	25
3	36	18	36
4	49	25	49

33. Completing the table:

a	b	$(a+b)^2$	a^2+b^2	a^2+ab+b^2	$a^2+2ab+b^2$
1	1	4	2	3	4
3	5	64	34	49	64
3	4	49	25	37	49
4	5	81	41	61	81

35. Multiplying using the FOIL method: $(a+5)(a-5)=a^2+5a-5a-25=a^2-25$

37. Multiplying using the FOIL method: $(y-1)(y+1)=y^2-y+y-1=y^2-1$

39. Multiplying using the difference of squares formula: $(9+x)(9-x)=(9)^2-(x)^2=81-x^2$

41. Multiplying using the difference of squares formula: $(2x+5)(2x-5)=(2x)^2-(5)^2=4x^2-25$

43. Multiplying using the difference of squares formula: $\left(4x+\frac{1}{3}\right)\left(4x-\frac{1}{3}\right)=(4x)^2-\left(\frac{1}{3}\right)^2=16x^2-\frac{1}{9}$

45. Multiplying using the difference of squares formula: $(2a+7)(2a-7)=(2a)^2-(7)^2=4a^2-49$

47. Multiplying using the difference of squares formula: $(6-7x)(6+7x)=(6)^2-(7x)^2=36-49x^2$

49. Multiplying using the difference of squares formula: $\left(x^2+3\right)\left(x^2-3\right)=\left(x^2\right)^2-(3)^2=x^4-9$

51. Multiplying using the difference of squares formula: $\left(a^2+4\right)\left(a^2-4\right)=\left(a^2\right)^2-(4)^2=a^4-16$

53. Multiplying using the difference of squares formula: $\left(5y^4-8\right)\left(5y^4+8\right)=\left(5y^4\right)^2-(8)^2=25y^8-64$

55. Multiplying and simplifying: $(x+3)(x-3)+(x+5)(x-5)=\left(x^2-9\right)+\left(x^2-25\right)=2x^2-34$

57. Multiplying and simplifying:
$$(2x+3)^2-(4x-1)^2=\left(4x^2+12x+9\right)-\left(16x^2-8x+1\right)=4x^2+12x+9-16x^2+8x-1=-12x^2+20x+8$$

59. Multiplying and simplifying:
$$(a+1)^2-(a+2)^2+(a+3)^2=\left(a^2+2a+1\right)-\left(a^2+4a+4\right)+\left(a^2+6a+9\right)$$
$$=a^2+2a+1-a^2-4a-4+a^2+6a+9$$
$$=a^2+4a+6$$

61. Multiplying and simplifying:
$$(2x+3)^3=(2x+3)(2x+3)^2$$
$$=(2x+3)\left(4x^2+12x+9\right)$$
$$=8x^3+24x^2+18x+12x^2+36x+27$$
$$=8x^3+36x^2+54x+27$$

63. Finding the product: $49(51)=(50-1)(50+1)=(50)^2-(1)^2=2,500-1=2,499$

65. Evaluating when $x=2$:
$$(x+3)^2=(2+3)^2=(5)^2=25$$
$$x^2+6x+9=(2)^2+6(2)+9=4+12+9=25$$

67. Let x and $x+1$ represent the two integers. The expression can be written as:
$$(x)^2+(x+1)^2=x^2+\left(x^2+2x+1\right)=2x^2+2x+1$$

69. Let x, $x+1$, and $x+2$ represent the three integers. The expression can be written as:
$$(x)^2+(x+1)^2+(x+2)^2=x^2+\left(x^2+2x+1\right)+\left(x^2+4x+4\right)=3x^2+6x+5$$

71. Verifying the areas: $(a+b)^2=a^2+ab+ab+b^2=a^2+2ab+b^2$

73. The product is $(2x+1)(2x+1) = 4x^2 + 4x + 1$:

	x	x	1
x	x^2	x^2	x
x	x^2	x^2	x
1	x	x	1

75. Simplifying: $\dfrac{15x^2 y}{3xy} = \dfrac{15}{3} \cdot \dfrac{x^2}{x} \cdot \dfrac{y}{y} = 5x$

77. Simplifying: $\dfrac{35a^6 b^8}{70a^2 b^{10}} = \dfrac{35}{70} \cdot \dfrac{a^6}{a^2} \cdot \dfrac{b^8}{b^{10}} = \dfrac{1}{2} \cdot a^4 \cdot \dfrac{1}{b^2} = \dfrac{a^4}{2b^2}$

79. Graphing both lines: **81.** Graphing both lines:

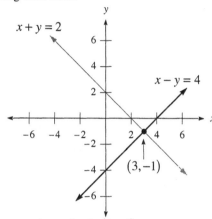
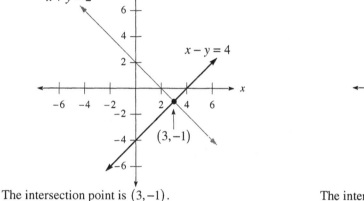

The intersection point is $(3, -1)$.   The intersection point is $(-1, 1)$.

5.7 Dividing a Polynomial by a Monomial

1. Performing the division: $\dfrac{5x^2 - 10x}{5x} = \dfrac{5x^2}{5x} - \dfrac{10x}{5x} = x - 2$

3. Performing the division: $\dfrac{15x - 10x^3}{5x} = \dfrac{15x}{5x} - \dfrac{10x^3}{5x} = 3 - 2x^2$

5. Performing the division: $\dfrac{25x^2 y - 10xy}{5x} = \dfrac{25x^2 y}{5x} - \dfrac{10xy}{5x} = 5xy - 2y$

7. Performing the division: $\dfrac{35x^5 - 30x^4 + 25x^3}{5x} = \dfrac{35x^5}{5x} - \dfrac{30x^4}{5x} + \dfrac{25x^3}{5x} = 7x^4 - 6x^3 + 5x^2$

9. Performing the division: $\dfrac{50x^5 - 25x^3 + 5x}{5x} = \dfrac{50x^5}{5x} - \dfrac{25x^3}{5x} + \dfrac{5x}{5x} = 10x^4 - 5x^2 + 1$

11. Performing the division: $\dfrac{8a^2 - 4a}{-2a} = \dfrac{8a^2}{-2a} + \dfrac{-4a}{-2a} = -4a + 2$

13. Performing the division: $\dfrac{16a^5 + 24a^4}{-2a} = \dfrac{16a^5}{-2a} + \dfrac{24a^4}{-2a} = -8a^4 - 12a^3$

15. Performing the division: $\dfrac{8ab + 10a^2}{-2a} = \dfrac{8ab}{-2a} + \dfrac{10a^2}{-2a} = -4b - 5a$

17. Performing the division: $\dfrac{12a^3 b - 6a^2 b^2 + 14ab^3}{-2a} = \dfrac{12a^3 b}{-2a} + \dfrac{-6a^2 b^2}{-2a} + \dfrac{14ab^3}{-2a} = -6a^2 b + 3ab^2 - 7b^3$

19. Performing the division: $\dfrac{a^2 + 2ab + b^2}{-2a} = \dfrac{a^2}{-2a} + \dfrac{2ab}{-2a} + \dfrac{b^2}{-2a} = -\dfrac{a}{2} - b - \dfrac{b^2}{2a}$

21. Performing the division: $\dfrac{6x + 8y}{2} = \dfrac{6x}{2} + \dfrac{8y}{2} = 3x + 4y$

23. Performing the division: $\dfrac{7y-21}{-7} = \dfrac{7y}{-7} + \dfrac{-21}{-7} = -y+3$

25. Performing the division: $\dfrac{10xy-8x}{2x} = \dfrac{10xy}{2x} - \dfrac{8x}{2x} = 5y-4$

27. Performing the division: $\dfrac{x^2y-x^3y^2}{x} = \dfrac{x^2y}{x} - \dfrac{x^3y^2}{x} = xy - x^2y^2$

29. Performing the division: $\dfrac{x^2y-x^3y^2}{-x^2y} = \dfrac{x^2y}{-x^2y} + \dfrac{-x^3y^2}{-x^2y} = -1+xy$

31. Performing the division: $\dfrac{a^2b^2-ab^2}{-ab^2} = \dfrac{a^2b^2}{-ab^2} + \dfrac{-ab^2}{-ab^2} = -a+1$

33. Performing the division: $\dfrac{x^3-3x^2y+xy^2}{x} = \dfrac{x^3}{x} - \dfrac{3x^2y}{x} + \dfrac{xy^2}{x} = x^2 - 3xy + y^2$

35. Performing the division: $\dfrac{10a^2-15a^2b+25a^2b^2}{5a^2} = \dfrac{10a^2}{5a^2} - \dfrac{15a^2b}{5a^2} + \dfrac{25a^2b^2}{5a^2} = 2-3b+5b^2$

37. Performing the division: $\dfrac{26x^2y^2-13xy}{-13xy} = \dfrac{26x^2y^2}{-13xy} + \dfrac{-13xy}{-13xy} = -2xy+1$

39. Performing the division: $\dfrac{4x^2y^2-2xy}{4xy} = \dfrac{4x^2y^2}{4xy} - \dfrac{2xy}{4xy} = xy - \dfrac{1}{2}$

41. Performing the division: $\dfrac{5a^2x-10ax^2+15a^2x^2}{20a^2x^2} = \dfrac{5a^2x}{20a^2x^2} - \dfrac{10ax^2}{20a^2x^2} + \dfrac{15a^2x^2}{20a^2x^2} = \dfrac{1}{4x} - \dfrac{1}{2a} + \dfrac{3}{4}$

43. Performing the division: $\dfrac{16x^5+8x^2+12x}{12x^3} = \dfrac{16x^5}{12x^3} + \dfrac{8x^2}{12x^3} + \dfrac{12x}{12x^3} = \dfrac{4x^2}{3} + \dfrac{2}{3x} + \dfrac{1}{x^2}$

45. Performing the division: $\dfrac{9a^{5m}-27a^{3m}}{3a^{2m}} = \dfrac{9a^{5m}}{3a^{2m}} - \dfrac{27a^{3m}}{3a^{2m}} = 3a^{5m-2m} - 9a^{3m-2m} = 3a^{3m} - 9a^m$

47. Performing the division:
$\dfrac{10x^{5m}-25x^{3m}+35x^m}{5x^m} = \dfrac{10x^{5m}}{5x^m} - \dfrac{25x^{3m}}{5x^m} + \dfrac{35x^m}{5x^m} = 2x^{5m-m} - 5x^{3m-m} + 7x^{m-m} = 2x^{4m} - 5x^{2m} + 7$

49. Simplifying and then dividing:
$$\dfrac{2x^3(3x+2)-3x^2(2x-4)}{2x^2} = \dfrac{6x^4+4x^3-6x^3+12x^2}{2x^2}$$
$$= \dfrac{6x^4-2x^3+12x^2}{2x^2}$$
$$= \dfrac{6x^4}{2x^2} - \dfrac{2x^3}{2x^2} + \dfrac{12x^2}{2x^2}$$
$$= 3x^2 - x + 6$$

51. Simplifying and then dividing:
$$\dfrac{(x+2)^2-(x-2)^2}{2x} = \dfrac{(x^2+4x+4)-(x^2-4x+4)}{2x} = \dfrac{x^2+4x+4-x^2+4x-4}{2x} = \dfrac{8x}{2x} = 4$$

53. Simplifying and then dividing:
$$\dfrac{(x+5)^2+(x+5)(x-5)}{2x} = \dfrac{(x^2+10x+25)+(x^2-25)}{2x} = \dfrac{2x^2+10x}{2x} = \dfrac{2x^2}{2x} + \dfrac{10x}{2x} = x+5$$

55. Evaluating each expression when $x=2$:
$$\dfrac{10x+15}{5} = \dfrac{10(2)+15}{5} = \dfrac{20+15}{5} = \dfrac{35}{5} = 7 \qquad 2x+3 = 2(2)+3 = 4+3 = 7$$

57. Evaluating each expression when $x = 10$:

$$\frac{3x+8}{2} = \frac{3(10)+8}{2} = \frac{30+8}{2} = \frac{38}{2} = 19 \qquad 3x+4 = 3(10)+4 = 30+4 = 34$$

Thus $\frac{3x+8}{2} \neq 3x+4$.

59. Adding the two equations:

$$2x = 14$$
$$x = 7$$

Substituting into the first equation:

$$7+y = 6$$
$$y = -1$$

The solution is $(7,-1)$.

61. Multiplying the second equation by 3:

$$2x-3y = -5$$
$$3x+3y = 15$$

Adding the two equations:

$$5x = 10$$
$$x = 2$$

Substituting into the second equation:

$$2+y = 5$$
$$y = 3$$

The solution is $(2,3)$.

63. Substituting into the first equation:

$$x+2x-1 = 2$$
$$3x-1 = 2$$
$$3x = 3$$
$$x = 1$$

Substituting into the second equation: $y = 2(1)-1 = 2-1 = 1$. The solution is $(1,1)$.

65. Substituting into the first equation:

$$4x+2(-2x+4) = 8$$
$$4x-4x+8 = 8$$
$$8 = 8$$

Since this statement is true, the system is dependent. The two lines coincide.

5.8 Dividing a Polynomial by a Polynomial

1. Using long division:

$$\begin{array}{r} x-2 \\ x-3\overline{)x^2-5x+6} \\ \underline{x^2-3x} \\ -2x+6 \\ \underline{-2x+6} \\ 0 \end{array}$$

The quotient is $x-2$.

3. Using long division:

$$\begin{array}{r} a+4 \\ a+5\overline{)a^2+9a+20} \\ \underline{a^2+5a} \\ 4a+20 \\ \underline{4a+20} \\ 0 \end{array}$$

The quotient is $a+4$.

5. Using long division:

$$\begin{array}{r} x-3 \\ x-3\overline{)x^2-6x+9} \\ \underline{x^2-3x} \\ -3x+9 \\ \underline{-3x+9} \\ 0 \end{array}$$

The quotient is $x-3$.

7. Using long division:

$$\begin{array}{r} x+3 \\ 2x-1\overline{)2x^2+5x-3} \\ \underline{2x^2-x} \\ 6x-3 \\ \underline{6x-3} \\ 0 \end{array}$$

The quotient is $x+3$.

9. Using long division:

$$2a+1 \overline{)\begin{array}{l} a-5 \\ 2a^2 - 9a - 5 \end{array}}$$

$$\begin{array}{l} \underline{2a^2 + \ a} \\ -10a - 5 \\ \underline{-10a - 5} \\ 0 \end{array}$$

The quotient is $a - 5$.

13. Using long division:

$$a+5 \overline{)\begin{array}{l} a-2 \\ a^2 + 3a + 2 \end{array}}$$

$$\begin{array}{l} \underline{a^2 + 5a} \\ -2a + 2 \\ \underline{-2a - 10} \\ 12 \end{array}$$

The quotient is $a - 2 + \dfrac{12}{a+5}$.

17. Using long division:

$$x+1 \overline{)\begin{array}{l} x+4 \\ x^2 + 5x - 6 \end{array}}$$

$$\begin{array}{l} \underline{x^2 + \ x} \\ 4x - 6 \\ \underline{4x + 4} \\ -10 \end{array}$$

The quotient is $x + 4 + \dfrac{-10}{x+1}$.

21. Using long division:

$$2x+4 \overline{)\begin{array}{l} x-3 \\ 2x^2 - 2x + 5 \end{array}}$$

$$\begin{array}{l} \underline{2x^2 + 4x} \\ -6x + 5 \\ \underline{-6x - 12} \\ 17 \end{array}$$

The quotient is $x - 3 + \dfrac{17}{2x+4}$.

25. Using long division:

$$3a-5 \overline{)\begin{array}{l} 2a^2 - a - 3 \\ 6a^3 - 13a^2 - 4a + 15 \end{array}}$$

$$\begin{array}{l} \underline{6a^3 - 10a^2} \\ -3a^2 - 4a \\ \underline{-3a^2 + 5a} \\ -9a + 15 \\ \underline{-9a + 15} \\ 0 \end{array}$$

The quotient is $2a^2 - a - 3$.

11. Using long division:

$$x+3 \overline{)\begin{array}{l} x+2 \\ x^2 + 5x + 8 \end{array}}$$

$$\begin{array}{l} \underline{x^2 + 3x} \\ 2x + 8 \\ \underline{2x + 6} \\ 2 \end{array}$$

The quotient is $x + 2 + \dfrac{2}{x+3}$.

15. Using long division:

$$x-2 \overline{)\begin{array}{l} x+4 \\ x^2 + 2x + 1 \end{array}}$$

$$\begin{array}{l} \underline{x^2 - 2x} \\ 4x + 1 \\ \underline{4x - 8} \\ 9 \end{array}$$

The quotient is $x + 4 + \dfrac{9}{x-2}$.

19. Using long division:

$$a+2 \overline{)\begin{array}{l} a+1 \\ a^2 + 3a + 1 \end{array}}$$

$$\begin{array}{l} \underline{a^2 + 2a} \\ a + 1 \\ \underline{a + 2} \\ -1 \end{array}$$

The quotient is $a + 1 + \dfrac{-1}{a+2}$.

23. Using long division:

$$2a+3 \overline{)\begin{array}{l} 3a-2 \\ 6a^2 + 5a + 1 \end{array}}$$

$$\begin{array}{l} \underline{6a^2 + 9a} \\ -4a + 1 \\ \underline{-4a - 6} \\ 7 \end{array}$$

The quotient is $3a - 2 + \dfrac{7}{2a+3}$.

27. Using long division:

$$x+1 \overline{)\begin{array}{l} x^2 - x + 5 \\ x^3 + 0x^2 + 4x + 5 \end{array}}$$

$$\begin{array}{l} \underline{x^3 + \ x^2} \\ -x^2 + 4x \\ \underline{-x^2 - \ x} \\ 5x + 5 \\ \underline{5x + 5} \\ 0 \end{array}$$

The quotient is $x^2 - x + 5$.

29. Using long division:

$$\begin{array}{r} x^2 + x + 1 \\ x - 1\overline{)\,x^3 + 0x^2 + 0x - 1} \\ \underline{x^3 - x^2} \\ x^2 + 0x \\ \underline{x^2 - x} \\ x - 1 \\ \underline{x - 1} \\ 0 \end{array}$$

The quotient is $x^2 + x + 1$.

31. Using long division:

$$\begin{array}{r} x^2 + 2x + 4 \\ x - 2\overline{)\,x^3 + 0x^2 + 0x - 8} \\ \underline{x^3 - 2x^2} \\ 2x^2 + 0x \\ \underline{2x^2 - 4x} \\ 4x - 8 \\ \underline{4x - 8} \\ 0 \end{array}$$

The quotient is $x^2 + 2x + 4$.

33. Let x and y represent the two numbers. The system of equations is:
$$x + y = 25$$
$$y = 4x$$
Substituting into the first equation:
$$x + 4x = 25$$
$$5x = 25$$
$$x = 5$$
$$y = 4(5) = 20$$
The two numbers are 5 and 20.

35. Let x represent the amount invested at 8% and y represent the amount invested at 9%. The system of equations is:
$$x + y = 1200$$
$$0.08x + 0.09y = 100$$
Multiplying the first equation by –0.08:
$$-0.08x - 0.08y = -96$$
$$0.08x + 0.09y = 100$$
Adding the two equations:
$$0.01y = 4$$
$$y = 400$$
Substituting into the first equation:
$$x + 400 = 1200$$
$$x = 800$$
You have $800 invested at 8% and $400 invested at 9%.

37. Let x represent the number of $5 bills and $x + 4$ represent the number of $10 bills. The equation is:
$$5(x) + 10(x + 4) = 160$$
$$5x + 10x + 40 = 160$$
$$15x + 40 = 160$$
$$15x = 120$$
$$x = 8$$
$$x + 4 = 12$$
You have 8 $5 bills and 12 $10 bills.

39. Let x represent the gallons of 20% antifreeze and y represent the gallons of 60% antifreeze. The system of equations is:
$$x + y = 16$$
$$0.20x + 0.60y = 0.35(16)$$
Multiplying the first equation by –0.20:
$$-0.20x - 0.20y = -3.2$$
$$0.20x + 0.60y = 5.6$$
Adding the two equations:
$$0.40y = 2.4$$
$$y = 6$$
Substituting into the first equation:
$$x + 6 = 16$$
$$x = 10$$
The mixture contains 10 gallons of 20% antifreeze and 6 gallons of 60% antifreeze.

Chapter 5 Review

1. Simplifying the expression: $(-1)^3 = (-1)(-1)(-1) = -1$

3. Simplifying the expression: $\left(\frac{3}{7}\right)^2 = \left(\frac{3}{7}\right)\cdot\left(\frac{3}{7}\right) = \frac{9}{49}$

5. Simplifying the expression: $x^{15}\cdot x^7\cdot x^5\cdot x^3 = x^{15+7+5+3} = x^{30}$

7. Simplifying the expression: $\left(2^6\right)^4 = 2^{6\cdot4} = 2^{24}$

9. Simplifying the expression: $(-2xyz)^3 = (-2)^3\cdot x^3y^3z^3 = -8x^3y^3z^3$

11. Writing with positive exponents: $4x^{-5} = 4\cdot\dfrac{1}{x^5} = \dfrac{4}{x^5}$

13. Simplifying the expression: $\dfrac{a^9}{a^3} = a^{9-3} = a^6$

15. Simplifying the expression: $\dfrac{x^9}{x^{-6}} = x^{9-(-6)} = x^{9+6} = x^{15}$

17. Simplifying the expression: $(-3xy)^0 = 1$

19. Simplifying the expression: $\left(3x^3y^2\right)^2 = 3^2\,x^6y^4 = 9x^6y^4$

21. Simplifying the expression: $\left(-3xy^2\right)^{-3} = \dfrac{1}{\left(-3xy^2\right)^3} = \dfrac{1}{(-3)^3\,x^3y^6} = -\dfrac{1}{27x^3y^6}$

23. Simplifying the expression: $\dfrac{\left(x^{-3}\right)^3\left(x^6\right)^{-1}}{\left(x^{-5}\right)^{-4}} = \dfrac{x^{-9}\cdot x^{-6}}{x^{20}} = \dfrac{x^{-15}}{x^{20}} = x^{-15-20} = x^{-35} = \dfrac{1}{x^{35}}$

25. Simplifying the expression: $\dfrac{\left(10x^3y^5\right)\left(21x^2y^6\right)}{\left(7xy^3\right)\left(5x^9y\right)} = \dfrac{210x^5y^{11}}{35x^{10}y^4} = \dfrac{210}{35}\cdot\dfrac{x^5}{x^{10}}\cdot\dfrac{y^{11}}{y^4} = 6\cdot\dfrac{1}{x^5}\cdot y^7 = \dfrac{6y^7}{x^5}$

27. Simplifying the expression: $\dfrac{8x^8y^3}{2x^3y} - \dfrac{10x^6y^9}{5xy^7} = 4x^5y^2 - 2x^5y^2 = 2x^5y^2$

29. Finding the quotient: $\dfrac{4.6\times10^5}{2\times10^{-3}} = 2.3\times10^8$

31. Performing the operations: $\left(3a^2 - 5a + 5\right) + \left(5a^2 - 7a - 8\right) = \left(3a^2 + 5a^2\right) + (-5a - 7a) + (5 - 8) = 8a^2 - 12a - 3$

33. Performing the operations:
$$\left(4x^2 - 3x - 2\right) - \left(8x^2 + 3x - 2\right) = 4x^2 - 3x - 2 - 8x^2 - 3x + 2$$
$$= \left(4x^2 - 8x^2\right) + (-3x - 3x) + (-2 + 2)$$
$$= -4x^2 - 6x$$

35. Multiplying: $3x(4x - 7) = 3x(4x) - 3x(7) = 12x^2 - 21x$

37. Multiplying using the column method:
$$
\begin{array}{r}
a^2 + 5a - 4 \\
a + 1 \\
\hline
a^3 + 5a^2 - 4a \\
a^2 + 5a - 4 \\
\hline
a^3 + 6a^2 + a - 4
\end{array}
$$

39. Multiplying using the FOIL method: $(3x - 7)(2x - 5) = 6x^2 - 14x - 15x + 35 = 6x^2 - 29x + 35$

41. Multiplying using the difference of squares formula: $\left(a^2 - 3\right)\left(a^2 + 3\right) = \left(a^2\right)^2 - (3)^2 = a^4 - 9$

43. Multiplying using the square of binomial formula: $(3x+4)^2 = (3x)^2 + 2(3x)(4) + (4)^2 = 9x^2 + 24x + 16$

45. Performing the division: $\dfrac{10ab + 20a^2}{-5a} = \dfrac{10ab}{-5a} + \dfrac{20a^2}{-5a} = -2b - 4a$

47. Using long division:

$$
\begin{array}{r}
x + 9 \\
x+6 \overline{\smash{\big)}\, x^2 + 15x + 54} \\
\underline{x^2 + 6x} \\
9x + 54 \\
\underline{9x + 54} \\
0
\end{array}
$$

The quotient is $x + 9$.

49. Using long division:

$$
\begin{array}{r}
x^2 - 4x + 16 \\
x+4 \overline{\smash{\big)}\, x^3 + 0x^2 + 0x + 64} \\
\underline{x^3 + 4x^2} \\
-4x^2 + 0x \\
\underline{-4x^2 - 16x} \\
16x + 64 \\
\underline{16x + 64} \\
0
\end{array}
$$

The quotient is $x^2 - 4x + 16$.

51. Using long division:

$$
\begin{array}{r}
x^2 - 4x + 5 \\
2x+1 \overline{\smash{\big)}\, 2x^3 - 7x^2 + 6x + 10} \\
\underline{2x^3 + x^2} \\
-8x^2 + 6x \\
\underline{-8x^2 - 4x} \\
10x + 10 \\
\underline{10x + 5} \\
5
\end{array}
$$

The quotient is $x^2 - 4x + 5 + \dfrac{5}{2x+1}$.

Chapters 1-5 Cumulative Review

1. Simplifying the expression: $-\left(-\frac{3}{4}\right) = \frac{3}{4}$

3. Simplifying the expression: $6 \cdot 7 + 7 \cdot 9 = 42 + 63 = 105$

5. Simplifying the expression: $6(4a+2) - 3(5a-1) = 24a + 12 - 15a + 3 = 24a - 15a + 12 + 3 = 9a + 15$

7. Simplifying the expression: $-15 - (-3) = -15 + 3 = -12$

9. Simplifying the expression: $(-9)(-5) = 45$

11. Simplifying the expression: $(2y)^4 = 2^4 y^4 = 16y^4$

13. Simplifying the expression: $\dfrac{\left(12xy^5\right)^3 \left(16x^2 y^2\right)}{\left(8x^3 y^3\right)\left(3x^5 y\right)} = \dfrac{1728x^3 y^{15} \cdot 16x^2 y^2}{24x^8 y^4} = \dfrac{27648 x^5 y^{17}}{24x^8 y^4} = \dfrac{1152 y^{13}}{x^3}$

15. Multiplying using the square of binomial formula: $(5x-1)^2 = (5x)^2 - 2(5x)(1) + (1)^2 = 25x^2 - 10x + 1$

17. Multiplying using the column method:

$$
\begin{array}{r}
x^2 + x + 1 \\
x - 1 \\
\hline
x^3 + x^2 + x \\
-x^2 - x - 1 \\
\hline
x^3 - 1
\end{array}
$$

19. Solving the equation:

$$
\begin{aligned}
6a - 5 &= 4a \\
-5 &= -2a \\
a &= \tfrac{5}{2}
\end{aligned}
$$

21. Solving the equation:

$$
\begin{aligned}
2(3x+5) + 8 &= 2x + 10 \\
6x + 10 + 8 &= 2x + 10 \\
6x + 18 &= 2x + 10 \\
4x + 18 &= 10 \\
4x &= -8 \\
x &= -2
\end{aligned}
$$

23.   Solving the inequality:
$$3(2t-5)-7 \le 5(3t+1)+5$$
$$6t-15-7 \le 15t+5+5$$
$$6t-22 \le 15t+10$$
$$-9t-22 \le 10$$
$$-9t \le 32$$
$$t \ge -\tfrac{32}{9}$$

Graphing the solution set:

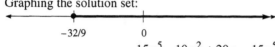

25.   Solving the compound inequality:
$$-5 \le 4x+3 \le 11$$
$$-8 \le 4x \le 8$$
$$-2 \le x \le 2$$

Graphing the solution set:

27.   Performing the division: $\dfrac{15x^5-10x^2+20x}{5x^5} = \dfrac{15x^5}{5x^5} - \dfrac{10x^2}{5x^5} + \dfrac{20x}{5x^5} = 3 - \dfrac{2}{x^3} + \dfrac{4}{x^4}$

28.   Graphing the equation:

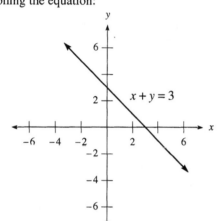

31.   Graphing the two lines:

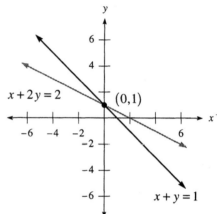

The intersection point is $(0,1)$.

33.   Graphing the two lines:

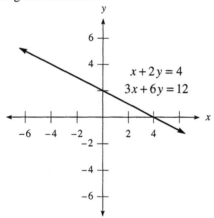

The system is dependent. The two lines coincide.

35.   Multiplying the first equation by –3:
$$-3x-6y=-15$$
$$3x+6y=14$$

Adding the two equations:
$$0=-1$$

Since this statement is false, there is no solution to the system. The two lines are parallel.

37.   To clear each equation of fractions, multiply the first equation by 12 and the second equation by 5:

$$12\left(\tfrac{1}{6}x + \tfrac{1}{4}y\right) = 12(1) \qquad\qquad 5\left(\tfrac{6}{5}x - y\right) = 5\left(\tfrac{8}{5}\right)$$
$$2x + 3y - 12 \qquad\qquad\qquad 6x - 5y = 8$$

The system of equations is:
$$2x + 3y = 12$$
$$6x - 5y = 8$$

Multiplying the first equation by -3:
$$-6x - 9y = -36$$
$$6x - 5y = 8$$

Adding the two equations:
$$-14y = -28$$
$$y = 2$$

Substituting into $2x + 3y = 12$:
$$2x + 3(2) = 12$$
$$2x + 6 = 12$$
$$2x = 6$$
$$x = 3$$

The solution is $(3,2)$.

39.   Solving the second equation for x:
$$2y = x - 3$$
$$x = 2y + 3$$

Substituting into the first equation:
$$4(2y + 3) + 5y = 25$$
$$8y + 12 + 5y = 25$$
$$13y + 12 = 25$$
$$13y = 13$$
$$y = 1$$
$$x = 2(1) + 3 = 5$$

The solution is $(5,1)$.

41.   Evaluating when $x = 3$: $8x - 3 = 8(3) - 3 = 24 - 3 = 21$

43.   The irrational numbers are $-\sqrt{2}$ and π.

45.   Substituting each ordered pair:

$(0,3)$: $3(0) - 4(3) = 0 - 12 = -12 \neq 12$

$(4,0)$: $3(4) - 4(0) = 12 - 0 = 12$

$\left(\tfrac{16}{3},1\right)$: $3\left(\tfrac{16}{3}\right) - 4(1) = 16 - 4 = 12$

The ordered pairs $(4,0)$ and $\left(\tfrac{16}{3},1\right)$ are solutions to the equation.

47.   First find the slope of the line: $m = \dfrac{-1-3}{6-(-2)} = \dfrac{-4}{8} = -\tfrac{1}{2}$. Using the point-slope formula:

$$y - 3 = -\tfrac{1}{2}\left(x - (-2)\right)$$
$$y - 3 = -\tfrac{1}{2}\left(x + 2\right)$$
$$y - 3 = -\tfrac{1}{2}x - 1$$
$$y = -\tfrac{1}{2}x + 2$$

To find the x-intercept, let $y = 0$:
$$0 = -\tfrac{1}{2}x + 2$$
$$-2 = -\tfrac{1}{2}x$$
$$x = 4$$

To find the y-intercept, let $x = 0$: $y = -\tfrac{1}{2}(0) + 2 = 0 + 2 = 2$

49. First find the slope: $m = \dfrac{8-4}{1-(-2)} = \dfrac{4}{3}$. Using the point-slope formula:

$$y - 8 = \tfrac{4}{3}(x-1)$$
$$y - 8 = \tfrac{4}{3}x - \tfrac{4}{3}$$
$$y = \tfrac{4}{3}x + \tfrac{20}{3}$$

Chapter 5 Test

1. Simplifying the expression: $(-3)^4 = (-3)(-3)(-3)(-3) = 81$

2. Simplifying the expression: $\left(\tfrac{3}{4}\right)^2 = \left(\tfrac{3}{4}\right)\left(\tfrac{3}{4}\right) = \tfrac{9}{16}$

3. Simplifying the expression: $\left(3x^3\right)^2 \left(2x^4\right)^3 = 9x^6 \cdot 8x^{12} = 72x^{18}$

4. Simplifying the expression: $3^{-2} = \dfrac{1}{3^2} = \tfrac{1}{9}$ **5.** Simplifying the expression: $\left(3a^4 b^2\right)^0 = 1$

6. Simplifying the expression: $\dfrac{a^{-3}}{a^{-5}} = a^{-3-(-5)} = a^{-3+5} = a^2$

7. Simplifying the expression: $\dfrac{\left(x^{-2}\right)^3 \left(x^{-3}\right)^{-5}}{\left(x^{-4}\right)^{-2}} = \dfrac{x^{-6}x^{15}}{x^8} = \dfrac{x^9}{x^8} = x^{9-8} = x$

8. Writing in scientific notation: $0.0278 = 2.78 \times 10^{-2}$ **9.** Writing in expanded form: $2.43 \times 10^5 = 243{,}000$

10. Simplifying the expression: $\dfrac{35x^2 y^4 z}{70 x^6 y^2 z} = \dfrac{35}{70} \cdot \dfrac{x^2}{x^6} \cdot \dfrac{y^4}{y^2} \cdot \dfrac{z}{z} = \dfrac{1}{2} \cdot \dfrac{1}{x^4} \cdot y^2 = \dfrac{y^2}{2x^4}$

11. Simplifying the expression: $\dfrac{\left(6a^2 b\right)\left(9a^3 b^2\right)}{18a^4 b^3} = \dfrac{54a^5 b^3}{18a^4 b^3} = 3a$

12. Simplifying the expression: $\dfrac{24x^7}{3x^2} + \dfrac{14x^9}{7x^4} = 8x^5 + 2x^5 = 10x^5$

13. Simplifying the expression: $\dfrac{\left(2.4 \times 10^5\right)\left(4.5 \times 10^{-2}\right)}{1.2 \times 10^{-6}} = \dfrac{10.8 \times 10^3}{1.2 \times 10^{-6}} = 9.0 \times 10^9$

14. Performing the operations: $8x^2 - 4x + 6x + 2 = 8x^2 + 2x + 2$

15. Performing the operations: $\left(5x^2 - 3x + 4\right) - \left(2x^2 - 7x - 2\right) = 5x^2 - 3x + 4 - 2x^2 + 7x + 2 = 3x^2 + 4x + 6$

16. Performing the operations: $(6x - 8) - (3x - 4) = 6x - 8 - 3x + 4 = 3x - 4$

17. Evaluating when $y = -2$: $2y^2 - 3y - 4 = 2(-2)^2 - 3(-2) - 4 = 2(4) + 6 - 4 = 8 + 6 - 4 = 10$

18. Multiplying using the distributive property:
$$2a^2 \left(3a^2 - 5a + 4\right) = 2a^2 \left(3a^2\right) - 2a^2 (5a) + 2a^2 (4) = 6a^4 - 10a^3 + 8a^2$$

19. Multiplying using the FOIL method: $\left(x + \tfrac{1}{2}\right)\left(x + \tfrac{1}{3}\right) = x^2 + \tfrac{1}{2}x + \tfrac{1}{3}x + \tfrac{1}{6} = x^2 + \tfrac{5}{6}x + \tfrac{1}{6}$

20. Multiplying using the FOIL method: $(4x - 5)(2x + 3) = 8x^2 - 10x + 12x - 15 = 8x^2 + 2x - 15$

21. Multiplying using the column method:

$$
\begin{array}{r}
x^2 + 3x + 9 \\
\underline{x - 3} \\
x^3 + 3x^2 + 9x \\
\underline{-3x^2 - 9x - 27} \\
x^3 - 27
\end{array}
$$

22.   Multiplying using the square of binomial formula: $(x+5)^2 = x^2 + 2(x)(5) + (5)^2 = x^2 + 10x + 25$

23.   Multiplying using the square of binomial formula: $(3a-2b)^2 = (3a)^2 - 2(3a)(2b) + (2b)^2 = 9a^2 - 12ab + 4b^2$

24.   Multiplying using the difference of squares formula: $(3x-4y)(3x+4y) = (3x)^2 - (4y)^2 = 9x^2 - 16y^2$

25.   Multiplying using the difference of squares formula: $(a^2-3)(a^2+3) = (a^2)^2 - (3)^2 = a^4 - 9$

26.   Performing the division: $\dfrac{10x^3 + 15x^2 - 5x}{5x} = \dfrac{10x^3}{5x} + \dfrac{15x^2}{5x} - \dfrac{5x}{5x} = 2x^2 + 3x - 1$

27.   Using long division:

$$
\begin{array}{r}
4x+3 \\
2x-3\overline{\smash{\big)}\,8x^2 - 6x - 5} \\
\underline{8x^2 - 12x} \\
6x - 5 \\
\underline{6x - 9} \\
4
\end{array}
$$

The quotient is $4x + 3 + \dfrac{4}{2x-3}$.

28.   Using long division:

$$
\begin{array}{r}
3x^2 + 9x + 25 \\
x-3\overline{\smash{\big)}\,3x^3 + 0x^2 - 2x + 1} \\
\underline{3x^3 - 9x^2} \\
9x^2 - 2x \\
\underline{9x^2 - 27x} \\
25x + 1 \\
\underline{25x - 75} \\
76
\end{array}
$$

The quotient is $3x^2 + 9x + 25 + \dfrac{76}{x-3}$.

29.   Using the volume formula: $V = (2.5 \text{ cm})^3 = 15.625 \text{ cm}^3$

30.   Let x represent the width, $5x$ represent the length, and $\frac{1}{5}x$ represent the height. The volume is given by:

$$V = (x)(5x)\left(\tfrac{1}{5}x\right) = x^3$$

Chapter 6
Factoring

6.1 The Greatest Common Factor and Factoring by Grouping

1. Factoring out the greatest common factor: $15x + 25 = 5(3x + 5)$

3. Factoring out the greatest common factor: $6a + 9 = 3(2a + 3)$

5. Factoring out the greatest common factor: $4x - 8y = 4(x - 2y)$

7. Factoring out the greatest common factor: $3x^2 - 6x - 9 = 3(x^2 - 2x - 3)$

9. Factoring out the greatest common factor: $3a^2 - 3a - 60 = 3(a^2 - a - 20)$

11. Factoring out the greatest common factor: $24y^2 - 52y + 24 = 4(6y^2 - 13y + 6)$

13. Factoring out the greatest common factor: $9x^2 - 8x^3 = x^2(9 - 8x)$

15. Factoring out the greatest common factor: $13a^2 - 26a^3 = 13a^2(1 - 2a)$

17. Factoring out the greatest common factor: $21x^2y - 28xy^2 = 7xy(3x - 4y)$

19. Factoring out the greatest common factor: $22a^2b^2 - 11ab^2 = 11ab^2(2a - 1)$

21. Factoring out the greatest common factor: $7x^3 + 21x^2 - 28x = 7x(x^2 + 3x - 4)$

23. Factoring out the greatest common factor: $121y^4 - 11x^4 = 11(11y^4 - x^4)$

25. Factoring out the greatest common factor: $100x^4 - 50x^3 + 25x^2 = 25x^2(4x^2 - 2x + 1)$

27. Factoring out the greatest common factor: $8a^2 + 16b^2 + 32c^2 = 8(a^2 + 2b^2 + 4c^2)$

29. Factoring out the greatest common factor: $4a^2b - 16ab^2 + 32a^2b^2 = 4ab(a - 4b + 8ab)$

31. Factoring out the greatest common factor: $121a^3b^2 - 22a^2b^3 + 33a^3b^3 = 11a^2b^2(11a - 2b + 3ab)$

33. Factoring out the greatest common factor: $12x^2y^3 - 72x^5y^3 - 36x^4y^4 = 12x^2y^3(1 - 6x^3 - 3x^2y)$

35. Factoring by grouping: $xy + 5x + 3y + 15 = x(y + 5) + 3(y + 5) = (y + 5)(x + 3)$

37. Factoring by grouping: $xy + 6x + 2y + 12 = x(y + 6) + 2(y + 6) = (y + 6)(x + 2)$

39. Factoring by grouping: $ab + 7a - 3b - 21 = a(b + 7) - 3(b + 7) = (b + 7)(a - 3)$

41. Factoring by grouping: $ax - bx + ay - by = x(a - b) + y(a - b) = (a - b)(x + y)$

43. Factoring by grouping: $2ax + 6x - 5a - 15 = 2x(a + 3) - 5(a + 3) = (a + 3)(2x - 5)$

45. Factoring by grouping: $3xb - 4b - 6x + 8 = b(3x - 4) - 2(3x - 4) = (3x - 4)(b - 2)$

47. Factoring by grouping: $x^2 + ax + 2x + 2a = x(x + a) + 2(x + a) = (x + a)(x + 2)$

49. Factoring by grouping: $x^2 - ax - bx + ab = x(x - a) - b(x - a) = (x - a)(x - b)$

51. Factoring by grouping: $ax + ay + bx + by + cx + cy = a(x+y) + b(x+y) + c(x+y) = (x+y)(a+b+c)$

53. Factoring by grouping: $6x^2 + 9x + 4x + 6 = 3x(2x+3) + 2(2x+3) = (2x+3)(3x+2)$

55. Factoring by grouping: $20x^2 - 2x + 50x - 5 = 2x(10x-1) + 5(10x-1) = (10x-1)(2x+5)$

57. Factoring by grouping: $20x^2 + 4x + 25x + 5 = 4x(5x+1) + 5(5x+1) = (5x+1)(4x+5)$

59. Factoring by grouping: $x^3 + 2x^2 + 3x + 6 = x^2(x+2) + 3(x+2) = (x+2)(x^2+3)$

61. Factoring by grouping: $6x^3 - 4x^2 + 15x - 10 = 2x^2(3x-2) + 5(3x-2) = (3x-2)(2x^2+5)$

63. Its greatest common factor is $3 \cdot 2 = 6$.

65. The correct factoring is: $12x^2 + 6x + 3 = 3(4x^2 + 2x + 1)$

67. The factored form is: $A = 1000 + 1000r = 1000(1+r)$
Substituting $r = 0.12$: $A = 1000(1+0.12) = 1000(1.12) = \$1,120$

69. **a.** Factoring: $A = 1,000,000 + 1,000,000r = 1,000,000(1+r)$
b. Substituting $r = 0.30$: $A = 1,000,000(1+0.3) = 1,300,000$ bacteria

71. Multiplying using the FOIL method: $(x-7)(x+2) = x^2 - 7x + 2x - 14 = x^2 - 5x - 14$

73. Multiplying using the FOIL method: $(x-3)(x+2) = x^2 - 3x + 2x - 6 = x^2 - x - 6$

75. Multiplying using the column method:

$$
\begin{array}{r}
x^2 - 3x + 9 \\
x + 3 \\
\hline
x^3 - 3x^2 + 9x \\
3x^2 - 9x + 27 \\
\hline
x^3 + 27
\end{array}
$$

77. Multiplying using the column method:

$$
\begin{array}{r}
x^2 + 4x - 3 \\
2x + 1 \\
\hline
2x^3 + 8x^2 - 6x \\
x^2 + 4x - 3 \\
\hline
2x^3 + 9x^2 - 2x - 3
\end{array}
$$

6.2 Factoring Trinomials

1. Factoring the trinomial: $x^2 + 7x + 12 = (x+3)(x+4)$ **3.** Factoring the trinomial: $x^2 + 3x + 2 = (x+2)(x+1)$

5. Factoring the trinomial: $a^2 + 10a + 21 = (a+7)(a+3)$ **7.** Factoring the trinomial: $x^2 - 7x + 10 = (x-5)(x-2)$

9. Factoring the trinomial: $y^2 - 10y + 21 = (y-7)(y-3)$ **11.** Factoring the trinomial: $x^2 - x - 12 = (x-4)(x+3)$

13. Factoring the trinomial: $y^2 + y - 12 = (y+4)(y-3)$ **15.** Factoring the trinomial: $x^2 + 5x - 14 = (x+7)(x-2)$

17. Factoring the trinomial: $r^2 - 8r - 9 = (r-9)(r+1)$ **19.** Factoring the trinomial: $x^2 - x - 30 = (x-6)(x+5)$

21. Factoring the trinomial: $a^2 + 15a + 56 = (a+7)(a+8)$ **23.** Factoring the trinomial: $y^2 - y - 42 = (y-7)(y+6)$

25. Factoring the trinomial: $x^2 + 13x + 42 = (x+7)(x+6)$

27. Factoring the trinomial: $2x^2 + 6x + 4 = 2(x^2 + 3x + 2) = 2(x+2)(x+1)$

29. Factoring the trinomial: $3a^2 - 3a - 60 = 3(a^2 - a - 20) = 3(a-5)(a+4)$

31. Factoring the trinomial: $100x^2 - 500x + 600 = 100(x^2 - 5x + 6) = 100(x-3)(x-2)$

33. Factoring the trinomial: $100p^2 - 1300p + 4000 = 100(p^2 - 13p + 40) = 100(p-8)(p-5)$

35. Factoring the trinomial: $x^4 - x^3 - 12x^2 = x^2(x^2 - x - 12) = x^2(x-4)(x+3)$

37. Factoring the trinomial: $2r^3 + 4r^2 - 30r = 2r(r^2 + 2r - 15) = 2r(r+5)(r-3)$

39. Factoring the trinomial: $2y^4 - 6y^3 - 8y^2 = 2y^2(y^2 - 3y - 4) = 2y^2(y-4)(y+1)$

41. Factoring the trinomial: $x^5 + 4x^4 + 4x^3 = x^3(x^2 + 4x + 4) = x^3(x+2)(x+2) = x^3(x+2)^2$

42. Factoring the trinomial: $x^5 + 13x^4 + 42x^3 = x^3(x^2 + 13x + 42) = x^3(x+7)(x+6)$

43. Factoring the trinomial: $3y^4 - 12y^3 - 15y^2 = 3y^2(y^2 - 4y - 5) = 3y^2(y-5)(y+1)$

45. Factoring the trinomial: $4x^4 - 52x^3 + 144x^2 = 4x^2(x^2 - 13x + 36) = 4x^2(x-9)(x-4)$

47. Factoring the trinomial: $x^2 + 5xy + 6y^2 = (x + 2y)(x + 3y)$

49. Factoring the trinomial: $x^2 - 9xy + 20y^2 = (x - 4y)(x - 5y)$

51. Factoring the trinomial: $a^2 + 2ab - 8b^2 = (a + 4b)(a - 2b)$

53. Factoring the trinomial: $a^2 - 10ab + 25b^2 = (a - 5b)(a - 5b) = (a - 5b)^2$

55. Factoring the trinomial: $a^2 + 10ab + 25b^2 = (a + 5b)(a + 5b) = (a + 5b)^2$

57. Factoring the trinomial: $x^2 + 2xa - 48a^2 = (x + 8a)(x - 6a)$

59. Factoring the trinomial: $x^2 - 5xb - 36b^2 = (x - 9b)(x + 4b)$

61. Factoring the trinomial: $x^4 - 5x^2 + 6 = \left(x^2 - 2\right)\left(x^2 - 3\right)$

63. Factoring the trinomial: $x^2 - 80x - 2000 = (x - 100)(x + 20)$

65. Factoring the trinomial: $x^2 - x + \frac{1}{4} = \left(x - \frac{1}{2}\right)\left(x - \frac{1}{2}\right) = \left(x - \frac{1}{2}\right)^2$

67. Factoring the trinomial: $x^2 + 0.6x + 0.08 = (x + 0.4)(x + 0.2)$

69. We can use long division to find the other factor:

$$\begin{array}{r} x + 250 \\ x + 10 \overline{)\, x^2 + 260x + 2500} \\ \underline{x^2 + 10x} \\ 250x + 2500 \\ \underline{250x + 2500} \\ 0 \end{array}$$

The other factor is $x + 250$.

71. Using FOIL to multiply out the factors: $(4x + 3)(x - 1) = 4x^2 + 3x - 4x - 3 = 4x^2 - x - 3$

73. Multiplying using the FOIL method: $(6a + 1)(a + 2) = 6a^2 + a + 12a + 2 = 6a^2 + 13a + 2$

75. Multiplying using the FOIL method: $(3a + 2)(2a + 1) = 6a^2 + 4a + 3a + 2 = 6a^2 + 7a + 2$

77. Multiplying using the FOIL method: $(6a + 2)(a + 1) = 6a^2 + 2a + 6a + 2 = 6a^2 + 8a + 2$

79. Subtracting the polynomials: $\left(5x^2 + 5x - 4\right) - \left(3x^2 - 2x + 7\right) = 5x^2 + 5x - 4 - 3x^2 + 2x - 7 = 2x^2 + 7x - 11$

81. Subtracting the polynomials: $(7x + 3) - (4x - 5) = 7x + 3 - 4x + 5 = 3x + 8$

83. Subtracting the polynomials: $\left(5x^2 - 5\right) - \left(2x^2 - 4x\right) = 5x^2 - 5 - 2x^2 + 4x = 3x^2 + 4x - 5$

6.3 More Trinomials to Factor

1. Factoring the trinomial: $2x^2 + 7x + 3 = (2x + 1)(x + 3)$

3. Factoring the trinomial: $2a^2 - a - 3 = (2a - 3)(a + 1)$

5. Factoring the trinomial: $3x^2 + 2x - 5 = (3x + 5)(x - 1)$

7. Factoring the trinomial: $3y^2 - 14y - 5 = (3y + 1)(y - 5)$

9. Factoring the trinomial: $6x^2 + 13x + 6 = (3x + 2)(2x + 3)$

11. Factoring the trinomial: $4x^2 - 12xy + 9y^2 = (2x - 3y)(2x - 3y) = (2x - 3y)^2$

13. Factoring the trinomial: $4y^2 - 11y - 3 = (4y + 1)(y - 3)$

15. Factoring the trinomial: $20x^2 - 41x + 20 = (4x - 5)(5x - 4)$

17. Factoring the trinomial: $20a^2 + 48ab - 5b^2 = (10a - b)(2a + 5b)$

19. Factoring the trinomial: $20x^2 - 21x - 5 = (4x - 5)(5x + 1)$

21. Factoring the trinomial: $12m^2 + 16m - 3 = (6m - 1)(2m + 3)$

23. Factoring the trinomial: $20x^2 + 37x + 15 = (4x + 5)(5x + 3)$

25. Factoring the trinomial: $12a^2 - 25ab + 12b^2 = (3a - 4b)(4a - 3b)$

27. Factoring the trinomial: $3x^2 - xy - 14y^2 = (3x - 7y)(x + 2y)$

29.   Factoring the trinomial: $14x^2 + 29x - 15 = (2x+5)(7x-3)$

31.   Factoring the trinomial: $6x^2 - 43x + 55 = (3x-5)(2x-11)$

33.   Factoring the trinomial: $15t^2 - 67t + 38 = (5t-19)(3t-2)$

35.   Factoring the trinomial: $4x^2 + 2x - 6 = 2(2x^2 + x - 3) = 2(2x+3)(x-1)$

37.   Factoring the trinomial: $24a^2 - 50a + 24 = 2(12a^2 - 25a + 12) = 2(4a-3)(3a-4)$

39.   Factoring the trinomial: $10x^3 - 23x^2 + 12x = x(10x^2 - 23x + 12) = x(5x-4)(2x-3)$

41.   Factoring the trinomial: $6x^4 - 11x^3 - 10x^2 = x^2(6x^2 - 11x - 10) = x^2(3x+2)(2x-5)$

43.   Factoring the trinomial: $10a^3 - 6a^2 - 4a = 2a(5a^2 - 3a - 2) = 2a(5a+2)(a-1)$

45.   Factoring the trinomial: $15x^3 - 102x^2 - 21x = 3x(5x^2 - 34x - 7) = 3x(5x+1)(x-7)$

47.   Factoring the trinomial: $35y^3 - 60y^2 - 20y = 5y(7y^2 - 12y - 4) = 5y(7y+2)(y-2)$

49.   Factoring the trinomial: $15a^4 - 2a^3 - a^2 = a^2(15a^2 - 2a - 1) = a^2(5a+1)(3a-1)$

51.   Factoring the trinomial: $24x^2y - 6xy - 45y = 3y(8x^2 - 2x - 15) = 3y(4x+5)(2x-3)$

53.   Factoring the trinomial: $12x^2y - 34xy^2 + 14y^3 = 2y(6x^2 - 17xy + 7y^2) = 2y(2x-y)(3x-7y)$

55.   Evaluating each expression when $x = 2$:
$$2x^2 + 7x + 3 = 2(2)^2 + 7(2) + 3 = 8 + 14 + 3 = 25$$
$$(2x+1)(x+3) = (2 \cdot 2 + 1)(2+3) = (5)(5) = 25$$

57.   Multiplying using the difference of squares formula: $(2x+3)(2x-3) = (2x)^2 - (3)^2 = 4x^2 - 9$

59.   Multiplying using the difference of squares formula:
$$(x+3)(x-3)(x^2+9) = (x^2 - 9)(x^2 + 9) = (x^2)^2 - (9)^2 = x^4 - 81$$

61.   Factoring: $h = 8 + 62t - 16t^2 = 2(4 + 31t - 8t^2) = 2(4-t)(1+8t)$. Now completing the table:

Time t (seconds)	Height h (feet)
0	8
1	54
2	68
3	50
4	0

63.   **a.**   Factoring: $V = 99x - 40x^2 + 4x^3 = x(99 - 40x + 4x^2) = x(9-2x)(11-2x)$

       b.   The original dimensions were 9 inches by 11 inches.

65.   Multiplying using the difference of squares formula: $(x+3)(x-3) = x^2 - (3)^2 = x^2 - 9$

67.   Multiplying using the difference of squares formula: $(6a+1)(6a-1) = (6a)^2 - (1)^2 = 36a^2 - 1$

69.   Multiplying using the square of binomial formula: $(x+4)^2 = x^2 + 2(x)(4) + (4)^2 = x^2 + 8x + 16$

71.   Multiplying using the square of binomial formula: $(2x+3)^2 = (2x)^2 + 2(2x)(3) + (3)^2 = 4x^2 + 12x + 9$

6.4 The Difference of Two Squares

1. Factoring the binomial: $x^2 - 9 = (x+3)(x-3)$

5. Factoring the binomial: $x^2 - 49 = (x+7)(x-7)$

7. Factoring the binomial: $4a^2 - 16 = 4(a^2 - 4) = 4(a+2)(a-2)$

9. The expression $9x^2 + 25$ cannot be factored.

11. Factoring the binomial: $25x^2 - 169 = (5x+13)(5x-13)$

13. Factoring the binomial: $9a^2 - 16b^2 = (3a+4b)(3a-4b)$

15. Factoring the binomial: $9 - m^2 = (3+m)(3-m)$

17. Factoring the binomial: $25 - 4x^2 = (5+2x)(5-2x)$

19. Factoring the binomial: $2x^2 - 18 = 2(x^2 - 9) = 2(x+3)(x-3)$

21. Factoring the binomial: $32a^2 - 128 = 32(a^2 - 4) = 32(a+2)(a-2)$

23. Factoring the binomial: $8x^2y - 18y = 2y(4x^2 - 9) = 2y(2x+3)(2x-3)$

25. Factoring the binomial: $a^4 - b^4 = (a^2 + b^2)(a^2 - b^2) = (a^2 + b^2)(a+b)(a-b)$

27. Factoring the binomial: $16m^4 - 81 = (4m^2 + 9)(4m^2 - 9) = (4m^2 + 9)(2m+3)(2m-3)$

29. Factoring the binomial: $3x^3y - 75xy^3 = 3xy(x^2 - 25y^2) = 3xy(x+5y)(x-5y)$

31. Factoring the trinomial: $x^2 - 2x + 1 = (x-1)(x-1) = (x-1)^2$

33. Factoring the trinomial: $x^2 + 2x + 1 = (x+1)(x+1) = (x+1)^2$

35. Factoring the trinomial: $a^2 - 10a + 25 = (a-5)(a-5) = (a-5)^2$

37. Factoring the trinomial: $y^2 + 4y + 4 = (y+2)(y+2) = (y+2)^2$

39. Factoring the trinomial: $x^2 - 4x + 4 = (x-2)(x-2) = (x-2)^2$

41. Factoring the trinomial: $m^2 - 12m + 36 = (m-6)(m-6) = (m-6)^2$

43. Factoring the trinomial: $4a^2 + 12a + 9 = (2a+3)(2a+3) = (2a+3)^2$

45. Factoring the trinomial: $49x^2 - 14x + 1 = (7x-1)(7x-1) = (7x-1)^2$

47. Factoring the trinomial: $9y^2 - 30y + 25 = (3y-5)(3y-5) = (3y-5)^2$

49. Factoring the trinomial: $x^2 + 10xy + 25y^2 = (x+5y)(x+5y) = (x+5y)^2$

51. Factoring the trinomial: $9a^2 + 6ab + b^2 = (3a+b)(3a+b) = (3a+b)^2$

53. Factoring the trinomial: $3a^2 + 18a + 27 = 3(a^2 + 6a + 9) = 3(a+3)(a+3) = 3(a+3)^2$

55. Factoring the trinomial: $2x^2 + 20xy + 50y^2 = 2(x^2 + 10xy + 25y^2) = 2(x+5y)(x+5y) = 2(x+5y)^2$

57. Factoring the trinomial: $5x^3 + 30x^2y + 45xy^2 = 5x(x^2 + 6xy + 9y^2) = 5x(x+3y)(x+3y) = 5x(x+3y)^2$

59. Factoring by grouping: $x^2 + 6x + 9 - y^2 = (x+3)^2 - y^2 = (x+3+y)(x+3-y)$

61. Factoring by grouping: $x^2 + 2xy + y^2 - 9 = (x+y)^2 - 9 = (x+y+3)(x+y-3)$

63. Since $(x+7)^2 = x^2 + 14x + 49$, the value is $b = 14$.

65. Since $(x+5)^2 = x^2 + 10x + 25$, the value is $c = 25$.

67.
    a. Subtracting square areas, the area is $x^2 - 4^2 = x^2 - 16$.
    b. Factoring: $x^2 - 16 = (x+4)(x-4)$
    c. The square can be rearranged to be a rectangle with dimensions $x - 4$ by $x + 4$. Cut off the right "flap", then place it along the bottom of the left rectangle to produce the desired rectangle.

69. The area is $a^2 - b^2 = (a+b)(a-b)$.

71. Using long division:

$$\begin{array}{r} x-2 \\ x-3{\overline{\smash{\big)}\,x^2-5x+8}} \end{array}$$

$$\begin{array}{r} x^2-3x \\ \hline -2x+8 \\ \underline{-2x+6} \\ 2 \end{array}$$

The quotient is $x-2+\dfrac{2}{x-3}$.

73. Using long division:

$$\begin{array}{r} 3x-2 \\ 2x+3{\overline{\smash{\big)}\,6x^2+5x+3}} \end{array}$$

$$\begin{array}{r} 6x^2+9x \\ \hline -4x+3 \\ \underline{-4x-6} \\ 9 \end{array}$$

The quotient is $3x-2+\dfrac{9}{2x+3}$.

6.5 Factoring: A General Review

1. Factoring the polynomial: $x^2-81=(x+9)(x-9)$

3. Factoring the polynomial: $x^2+2x-15=(x+5)(x-3)$

5. Factoring the polynomial: $x^2+6x+9=(x+3)(x+3)=(x+3)^2$

7. Factoring the polynomial: $y^2-10y+25=(y-5)(y-5)=(y-5)^2$

9. Factoring the polynomial: $2a^3b+6a^2b+2ab=2ab(a^2+3a+1)$

11. The polynomial x^2+x+1 cannot be factored.

13. Factoring the polynomial: $12a^2-75=3(4a^2-25)=3(2a+5)(2a-5)$

15. Factoring the polynomial: $9x^2-12xy+4y^2=(3x-2y)(3x-2y)=(3x-2y)^2$

17. Factoring the polynomial: $4x^3+16xy^2=4x(x^2+4y^2)$

19. Factoring the polynomial: $2y^3+20y^2+50y=2y(y^2+10y+25)=2y(y+5)(y+5)=2y(y+5)^2$

21. Factoring the polynomial: $a^6+4a^4b^2=a^4(a^2+4b^2)$

23. Factoring the polynomial: $xy+3x+4y+12=x(y+3)+4(y+3)=(y+3)(x+4)$

25. Factoring the polynomial: $x^4-16=(x^2+4)(x^2-4)=(x^2+4)(x+2)(x-2)$

27. Factoring the polynomial: $xy-5x+2y-10=x(y-5)+2(y-5)=(y-5)(x+2)$

29. Factoring the polynomial: $5a^2+10ab+5b^2=5(a^2+2ab+b^2)=5(a+b)(a+b)=5(a+b)^2$

31. The polynomial x^2+49 cannot be factored.

33. Factoring the polynomial: $3x^2+15xy+18y^2=3(x^2+5xy+6y^2)=3(x+2y)(x+3y)$

35. Factoring the polynomial: $2x^2+15x-38=(2x+19)(x-2)$

37. Factoring the polynomial: $100x^2-300x+200=100(x^2-3x+2)=100(x-2)(x-1)$

39. Factoring the polynomial: $x^2-64=(x+8)(x-8)$

41. Factoring the polynomial: $x^2+3x+ax+3a=x(x+3)+a(x+3)=(x+3)(x+a)$

43. Factoring the polynomial: $49a^7-9a^5=a^5(49a^2-9)=a^5(7a+3)(7a-3)$

45. The polynomial $49x^2+9y^2$ cannot be factored.

47. Factoring the polynomial: $25a^3+20a^2+3a=a(25a^2+20a+3)=a(5a+3)(5a+1)$

49. Factoring the polynomial: $xa-xb+ay-by=x(a-b)+y(a-b)=(a-b)(x+y)$

51. Factoring the polynomial: $48a^4b-3a^2b=3a^2b(16a^2-1)=3a^2b(4a+1)(4a-1)$

53. Factoring the polynomial: $20x^4-45x^2=5x^2(4x^2-9)=5x^2(2x+3)(2x-3)$

55. Factoring the polynomial: $3x^2+35xy-82y^2=(3x+41y)(x-2y)$

57. Factoring the polynomial: $16x^5-44x^4+30x^3=2x^3(8x^2-22x+15)=2x^3(2x-3)(4x-5)$

59. Factoring the polynomial: $2x^2 + 2ax + 3x + 3a = 2x(x+a) + 3(x+a) = (x+a)(2x+3)$

61. Factoring the polynomial: $y^4 - 1 = (y^2+1)(y^2-1) = (y^2+1)(y+1)(y-1)$

63. Factoring the polynomial:
$$12x^4y^2 + 36x^3y^3 + 27x^2y^4 = 3x^2y^2(4x^2 + 12xy + 9y^2) = 3x^2y^2(2x+3y)(2x+3y) = 3x^2y^2(2x+3y)^2$$

65. Solving the equation:
$$3x - 6 = 9$$
$$3x = 15$$
$$x = 5$$

67. Solving the equation:
$$2x + 3 = 0$$
$$2x = -3$$
$$x = -\frac{3}{2}$$

69. Solving the equation:
$$4x + 3 = 0$$
$$4x = -3$$
$$x = -\frac{3}{4}$$

71. Simplifying the expression: $x^8 \cdot x^7 = x^{8+7} = x^{15}$

73. Simplifying the expression: $\left(3x^3\right)^2\left(2x^4\right)^3 = 9x^6 \cdot 8x^{12} = 72x^{18}$

75. Writing in scientific notation: $57,600 = 5.76 \times 10^4$

6.6 Solving Equations by Factoring

1. Setting each factor equal to 0:
$$x + 2 = 0 \qquad\qquad x - 1 = 0$$
$$x = -2 \qquad\qquad x = 1$$
The solutions are –2 and 1.

3. Setting each factor equal to 0:
$$a - 4 = 0 \qquad\qquad a - 5 = 0$$
$$a = 4 \qquad\qquad a = 5$$
The solutions are 4 and 5.

5. Setting each factor equal to 0:
$$x = 0 \qquad x + 1 = 0 \qquad x - 3 = 0$$
$$x = -1 \qquad\qquad x = 3$$
The solutions are 0, –1 and 3.

7. Setting each factor equal to 0:
$$3x + 2 = 0 \qquad\qquad 2x + 3 = 0$$
$$3x = -2 \qquad\qquad 2x = -3$$
$$x = -\frac{2}{3} \qquad\qquad x = -\frac{3}{2}$$
The solutions are $-\frac{2}{3}$ and $-\frac{3}{2}$.

9. Setting each factor equal to 0:
$$m = 0 \qquad 3m + 4 = 0 \qquad 3m - 4 = 0$$
$$3m = -4 \qquad 3m = 4$$
$$m = -\frac{4}{3} \qquad m = \frac{4}{3}$$
The solutions are 0, $-\frac{4}{3}$ and $\frac{4}{3}$.

11. Setting each factor equal to 0:
$$2y = 0 \qquad 3y + 1 = 0 \qquad 5y + 3 = 0$$
$$y = 0 \qquad 3y = -1 \qquad 5y = -3$$
$$y = -\frac{1}{3} \qquad y = -\frac{3}{5}$$
The solutions are 0, $-\frac{1}{3}$ and $-\frac{3}{5}$.

13. Solving by factoring:
$$x^2 + 3x + 2 = 0$$
$$(x+2)(x+1) = 0$$
$$x = -2, -1$$

15. Solving by factoring:
$$x^2 - 9x + 20 = 0$$
$$(x-4)(x-5) = 0$$
$$x = 4, 5$$

17.   Solving by factoring:

$$a^2 - 2a - 24 = 0$$
$$(a-6)(a+4) = 0$$
$$a = 6, -4$$

19.   Solving by factoring:

$$100x^2 - 500x + 600 = 0$$
$$100\left(x^2 - 5x + 6\right) = 0$$
$$100(x-2)(x-3) = 0$$
$$x = 2, 3$$

21.   Solving by factoring:

$$x^2 = -6x - 9$$
$$x^2 + 6x + 9 = 0$$
$$(x+3)^2 = 0$$
$$x + 3 = 0$$
$$x = -3$$

23.   Solving by factoring:

$$a^2 - 16 = 0$$
$$(a+4)(a-4) = 0$$
$$a = -4, 4$$

25.   Solving by factoring:

$$2x^2 + 5x - 12 = 0$$
$$(2x-3)(x+4) = 0$$
$$x = \frac{3}{2}, -4$$

27.   Solving by factoring:

$$9x^2 + 12x + 4 = 0$$
$$(3x+2)^2 = 0$$
$$3x + 2 = 0$$
$$x = -\frac{2}{3}$$

29.   Solving by factoring:

$$a^2 + 25 = 10a$$
$$a^2 - 10a + 25 = 0$$
$$(a-5)^2 = 0$$
$$a - 5 = 0$$
$$a = 5$$

31.   Solving by factoring:

$$2x^2 = 3x + 20$$
$$2x^2 - 3x - 20 = 0$$
$$(2x+5)(x-4) = 0$$
$$x = -\frac{5}{2}, 4$$

33.   Solving by factoring:

$$3m^2 = 20 - 7m$$
$$3m^2 + 7m - 20 = 0$$
$$(3m-5)(m+4) = 0$$
$$m = \frac{5}{3}, -4$$

35.   Solving by factoring:

$$4x^2 - 49 = 0$$
$$(2x+7)(2x-7) = 0$$
$$x = -\frac{7}{2}, \frac{7}{2}$$

37.   Solving by factoring:

$$x^2 + 6x = 0$$
$$x(x+6) = 0$$
$$x = 0, -6$$

39.   Solving by factoring:

$$x^2 - 3x = 0$$
$$x(x-3) = 0$$
$$x = 0, 3$$

41.   Solving by factoring:

$$2x^2 = 8x$$
$$2x^2 - 8x = 0$$
$$2x(x-4) = 0$$
$$x = 0, 4$$

43.   Solving by factoring:

$$3x^2 = 15x$$
$$3x^2 - 15x = 0$$
$$3x(x-5) = 0$$
$$x = 0, 5$$

45.   Solving by factoring:

$$1400 = 400 + 700x - 100x^2$$
$$100x^2 - 700x + 1000 = 0$$
$$100\left(x^2 - 7x + 10\right) = 0$$
$$100(x-5)(x-2) = 0$$
$$x = 2, 5$$

47.   Solving by factoring:

$$6x^2 = -5x + 4$$
$$6x^2 + 5x - 4 = 0$$
$$(3x+4)(2x-1) = 0$$
$$x = -\frac{4}{3}, \frac{1}{2}$$

49. Solving by factoring:
$$x(2x-3)=20$$
$$2x^2-3x=20$$
$$2x^2-3x-20=0$$
$$(2x+5)(x-4)=0$$
$$x=-\tfrac{5}{2},4$$

51. Solving by factoring:
$$t(t+2)=80$$
$$t^2+2t=80$$
$$t^2+2t-80=0$$
$$(t+10)(t-8)=0$$
$$t=-10,8$$

53. Solving by factoring:
$$4000=(1300-100p)p$$
$$4000=1300p-100p^2$$
$$100p^2-1300p+4000=0$$
$$100\left(p^2-13p+40\right)=0$$
$$100(p-8)(p-5)=0$$
$$p=5,8$$

55. Solving by factoring:
$$x(14-x)=48$$
$$14x-x^2=48$$
$$-x^2+14x-48=0$$
$$x^2-14x+48=0$$
$$(x-6)(x-8)=0$$
$$x=6,8$$

57. Solving by factoring:
$$(x+5)^2=2x+9$$
$$x^2+10x+25=2x+9$$
$$x^2+8x+16=0$$
$$(x+4)^2=0$$
$$x+4=0$$
$$x=-4$$

59. Solving by factoring:
$$(y-6)^2=y-4$$
$$y^2-12y+36=y-4$$
$$y^2-13y+40=0$$
$$(y-5)(y-8)=0$$
$$y=5,8$$

61. Solving by factoring:
$$10^2=(x+2)^2+x^2$$
$$100=x^2+4x+4+x^2$$
$$100=2x^2+4x+4$$
$$0=2x^2+4x-96$$
$$0=2\left(x^2+2x-48\right)$$
$$0=2(x+8)(x-6)$$
$$x=-8,6$$

63. Solving by factoring:
$$2x^3+11x^2+12x=0$$
$$x\left(2x^2+11x+12\right)=0$$
$$x(2x+3)(x+4)=0$$
$$x=0,-\tfrac{3}{2},-4$$

65. Solving by factoring:
$$4y^3-2y^2-30y=0$$
$$2y\left(2y^2-y-15\right)=0$$
$$2y(2y+5)(y-3)=0$$
$$y=0,-\tfrac{5}{2},3$$

67. Solving by factoring:
$$8x^3+16x^2=10x$$
$$8x^3+16x^2-10x=0$$
$$2x\left(4x^2+8x-5\right)=0$$
$$2x(2x-1)(2x+5)=0$$
$$x=0,\tfrac{1}{2},-\tfrac{5}{2}$$

69. Solving by factoring:
$$20a^3=-18a^2+18a$$
$$20a^3+18a^2-18a=0$$
$$2a\left(10a^2+9a-9\right)=0$$
$$2a(5a-3)(2a+3)=0$$
$$a=0,\tfrac{3}{5},-\tfrac{3}{2}$$

71. Solving by factoring:
$$x^3+3x^2-4x-12=0$$
$$x^2(x+3)-4(x+3)=0$$
$$(x+3)\left(x^2-4\right)=0$$
$$(x+3)(x+2)(x-2)=0$$
$$x=-3,-2,2$$

73. Solving by factoring:
$$x^3 + x^2 - 16x - 16 = 0$$
$$x^2(x+1) - 16(x+1) = 0$$
$$(x+1)(x^2 - 16) = 0$$
$$(x+1)(x+4)(x-4) = 0$$
$$x = -1, -4, 4$$

75. Let x represent the cost of the suit and $5x$ represent the cost of the bicycle. The equation is:
$$x + 5x = 90$$
$$6x = 90$$
$$x = 15$$
$$5x = 75$$
The suit costs $15 and the bicycle costs $75.

77. Let x represent the cost of the lot and $4x$ represent the cost of the house. The equation is:
$$x + 4x = 3000$$
$$5x = 3000$$
$$x = 600$$
$$4x = 2400$$
The lot cost $600 and the house cost $2,400.

79. Simplifying using the properties of exponents: $2^{-3} = \dfrac{1}{2^3} = \dfrac{1}{8}$

81. Simplifying using the properties of exponents: $\dfrac{x^5}{x^{-3}} = x^{5-(-3)} = x^{5+3} = x^8$

83. Simplifying using the properties of exponents: $\dfrac{\left(x^2\right)^3}{\left(x^{-3}\right)^4} = \dfrac{x^6}{x^{-12}} = x^{6-(-12)} = x^{6+12} = x^{18}$

85. Writing in scientific notation: $0.0056 = 5.6 \times 10^{-3}$

6.7 Applications

1. Let x and $x + 2$ represent the two integers. The equation is:
$$x(x+2) = 80$$
$$x^2 + 2x = 80$$
$$x^2 + 2x - 80 = 0$$
$$(x+10)(x-8) = 0$$
$$x = -10, 8$$
$$x+2 = -8, 10$$
The two numbers are either -10 and -8, or 8 and 10.

3. Let x and $x + 2$ represent the two integers. The equation is:
$$x(x+2) = 99$$
$$x^2 + 2x = 99$$
$$x^2 + 2x - 99 = 0$$
$$(x+11)(x-9) = 0$$
$$x = -11, 9$$
$$x+2 = -9, 11$$
The two numbers are either -11 and -9, or 9 and 11.

5.   Let x and $x + 2$ represent the two integers. The equation is:

$$x(x+2) = 5(x+x+2) - 10$$
$$x^2 + 2x = 5(2x+2) - 10$$
$$x^2 + 2x = 10x + 10 - 10$$
$$x^2 + 2x = 10x$$
$$x^2 - 8x = 0$$
$$x(x-8) = 0$$
$$x = 0, 8$$
$$x + 2 = 2, 10$$

The two numbers are either 0 and 2, or 8 and 10.

7.   Let x and $14 - x$ represent the two numbers. The equation is:

$$x(14-x) = 48$$
$$14x - x^2 = 48$$
$$0 = x^2 - 14x + 48$$
$$0 = (x-8)(x-6)$$
$$x = 8, 6$$
$$14 - x = 6, 8$$

The two numbers are 6 and 8.

9.   Let x and $5x + 2$ represent the two numbers. The equation is:

$$x(5x+2) = 24$$
$$5x^2 + 2x = 24$$
$$5x^2 + 2x - 24 = 0$$
$$(5x+12)(x-2) = 0$$
$$x = -\tfrac{12}{5}, 2$$
$$5x + 2 = -10, 12$$

The two numbers are either $-\tfrac{12}{5}$ and -10, or 2 and 12.

11.   Let x and $4x$ represent the two numbers. The equation is:

$$x(4x) = 4(x+4x)$$
$$4x^2 = 4(5x)$$
$$4x^2 = 20x$$
$$4x^2 - 20x = 0$$
$$4x(x-5) = 0$$
$$x = 0, 5$$
$$4x = 0, 20$$

The two numbers are either 0 and 0, or 5 and 20.

13.   Let w represent the width and $w + 1$ represent the length. The equation is:

$$w(w+1) = 12$$
$$w^2 + w = 12$$
$$w^2 + w - 12 = 0$$
$$(w+4)(w-3) = 0$$
$$w = 3 \quad (w = -4 \text{ is impossible})$$
$$w + 1 = 4$$

The width is 3 inches and the length is 4 inches.

15.   Let b represent the base and $2b$ represent the height. The equation is:

$$\tfrac{1}{2}b(2b) = 9$$
$$b^2 = 9$$
$$b^2 - 9 = 0$$
$$(b+3)(b-3) = 0$$
$$b = 3 \qquad (b = -3 \text{ is impossible})$$

The base is 3 inches.

17.   Let x and $x + 2$ represent the two legs. The equation is:

$$x^2 + (x+2)^2 = 10^2$$
$$x^2 + x^2 + 4x + 4 = 100$$
$$2x^2 + 4x + 4 = 100$$
$$2x^2 + 4x - 96 = 0$$
$$2\left(x^2 + 2x - 48\right) = 0$$
$$2(x+8)(x-6) = 0$$
$$x = 6 \qquad (x = -8 \text{ is impossible})$$
$$x + 2 = 8$$

The legs are 6 inches and 8 inches.

19.   Let x represent the longer leg and $x + 1$ represent the hypotenuse. The equation is:

$$5^2 + x^2 = (x+1)^2$$
$$25 + x^2 = x^2 + 2x + 1$$
$$25 = 2x + 1$$
$$24 = 2x$$
$$x = 12$$

The longer leg is 12 meters.

21.   Setting $C = \$1,400$:

$$1400 = 400 + 700x - 100x^2$$
$$100x^2 - 700x + 1000 = 0$$
$$100\left(x^2 - 7x + 10\right) = 0$$
$$100(x-5)(x-2) = 0$$
$$x = 2, 5$$

The company can manufacture either 200 items or 500 items.

23.   Setting $C = \$2,200$:

$$2200 = 600 + 1000x - 100x^2$$
$$100x^2 - 1000x + 1600 = 0$$
$$100\left(x^2 - 10x + 16\right) = 0$$
$$100(x-2)(x-8) = 0$$
$$x = 2, 8$$

The company can manufacture either 200 videotapes or 800 videotapes.

25.   The revenue is given by: $R = xp = (1200 - 100p)p$. Setting $R = \$3,200$:

$$3200 = (1200 - 100p)p$$
$$3200 = 1200p - 100p^2$$
$$100p^2 - 1200p + 3200 = 0$$
$$100\left(p^2 - 12p + 32\right) = 0$$
$$100(p-4)(p-8) = 0$$
$$p = 4, 8$$

The company should sell the ribbons for either $4 or $8.

27.   The revenue is given by: $R = xp = (1700 - 100p)p$. Setting $R = \$7,000$:

$$7000 = (1700 - 100p)p$$
$$7000 = 1700p - 100p^2$$
$$100p^2 - 1700p + 7000 = 0$$
$$100(p^2 - 17p + 70) = 0$$
$$100(p - 7)(p - 10) = 0$$
$$p = 7, 10$$

The calculators should be sold for either \$7 or \$10.

29.   **a.**   Let x represent the distance from the base to the wall, and $2x + 2$ represent the height on the wall. Using the Pythagorean theorem:

$$x^2 + (2x + 2)^2 = 13^2$$
$$x^2 + 4x^2 + 8x + 4 = 169$$
$$5x^2 + 8x - 165 = 0$$
$$(5x + 33)(x - 5) = 0$$
$$x = 5 \quad \left(x = -\tfrac{33}{5} \text{ is impossible}\right)$$

The base of the ladder is 5 feet from the wall.

b.   Since $2x + 2 = 2 \cdot 5 + 2 = 12$, the ladder reaches a height of 12 feet.

31.   **a.**   Finding when $h = 0$:

$$0 = -16t^2 + 396t + 100$$
$$16t^2 - 396t - 100 = 0$$
$$4t^2 - 99t - 25 = 0$$
$$(4t + 1)(t - 25) = 0$$
$$t = 25 \quad \left(t = -\tfrac{1}{4} \text{ is impossible}\right)$$

The bullet will land on the ground after 25 seconds.

b.   Completing the table:

t (seconds)	h (feet)
0	100
5	1,680
10	2,460
15	2,440
20	1,620
25	0

33.   Simplifying the expression: $(5x^3)^2(2x^6)^3 = 25x^6 \cdot 8x^{18} = 200x^{24}$

35.   Simplifying the expression: $\dfrac{x^4}{x^{-3}} = x^{4-(-3)} = x^{4+3} = x^7$

37.   Simplifying the expression: $(2 \times 10^{-4})(4 \times 10^5) = 8 \times 10^1 = 80$

39.   Simplifying the expression: $20ab^2 - 16ab^2 + 6ab^2 = 10ab^2$

41.   Multiplying using the distributive property:
$$2x^2(3x^2 + 3x - 1) = 2x^2(3x^2) + 2x^2(3x) - 2x^2(1) = 6x^4 + 6x^3 - 2x^2$$

43.   Multiplying using the square of binomial formula: $(3y - 5)^2 = (3y)^2 - 2(3y)(5) + (5)^2 = 9y^2 - 30y + 25$

45.   Multiplying using the difference of squares formula: $(2a^2 + 7)(2a^2 - 7) = (2a^2)^2 - (7)^2 = 4a^4 - 49$

Chapter 6 Review

1. Factoring the polynomial: $10x - 20 = 10(x - 2)$
3. Factoring the polynomial: $5x - 5y = 5(x - y)$
5. Factoring the polynomial: $8x + 4 = 4(2x + 1)$
7. Factoring the polynomial: $24y^2 - 40y + 48 = 8\left(3y^2 - 5y + 6\right)$
9. Factoring the polynomial: $49a^3 - 14b^3 = 7\left(7a^3 - 2b^3\right)$
11. Factoring the polynomial: $xy + bx + ay + ab = x(y + b) + a(y + b) = (y + b)(x + a)$
13. Factoring the polynomial: $2xy + 10x - 3y - 15 = 2x(y + 5) - 3(y + 5) = (y + 5)(2x - 3)$
15. Factoring the polynomial: $y^2 + 9y + 14 = (y + 7)(y + 2)$
17. Factoring the polynomial: $a^2 - 14a + 48 = (a - 8)(a - 6)$
19. Factoring the polynomial: $y^2 + 20y + 99 = (y + 9)(y + 11)$
21. Factoring the polynomial: $2x^2 + 13x + 15 = (2x + 3)(x + 5)$
23. Factoring the polynomial: $5y^2 + 11y + 6 = (5y + 6)(y + 1)$
25. Factoring the polynomial: $6r^2 + 5rt - 6t^2 = (3r - 2t)(2r + 3t)$
27. Factoring the polynomial: $n^2 - 81 = (n + 9)(n - 9)$
29. The expression $x^2 + 49$ cannot be factored.
31. Factoring the polynomial: $64a^2 - 121b^2 = (8a + 11b)(8a - 11b)$
33. Factoring the polynomial: $y^2 + 20y + 100 = (y + 10)(y + 10) = (y + 10)^2$
35. Factoring the polynomial: $64t^2 + 16t + 1 = (8t + 1)(8t + 1) = (8t + 1)^2$
37. Factoring the polynomial: $4r^2 - 12rt + 9t^2 = (2r - 3t)(2r - 3t) = (2r - 3t)^2$
39. Factoring the polynomial: $2x^2 + 20x + 48 = 2\left(x^2 + 10x + 24\right) = 2(x + 4)(x + 6)$
41. Factoring the polynomial: $3m^3 - 18m^2 - 21m = 3m\left(m^2 - 6m - 7\right) = 3m(m - 7)(m + 1)$
43. Factoring the polynomial: $8x^2 + 16x + 6 = 2\left(4x^2 + 8x + 3\right) = 2(2x + 1)(2x + 3)$
45. Factoring the polynomial: $20m^3 - 34m^2 + 6m = 2m\left(10m^2 - 17m + 3\right) = 2m(5m - 1)(2m - 3)$
47. Factoring the polynomial: $4x^2 + 40x + 100 = 4\left(x^2 + 10x + 25\right) = 4(x + 5)(x + 5) = 4(x + 5)^2$
49. Factoring the polynomial: $5x^2 - 45 = 5\left(x^2 - 9\right) = 5(x + 3)(x - 3)$
51. Factoring the polynomial: $6a^3b + 33a^2b^2 + 15ab^3 = 3ab\left(2a^2 + 11ab + 5b^2\right) = 3ab(2a + b)(a + 5b)$
53. Factoring the polynomial: $4y^6 + 9y^4 = y^4\left(4y^2 + 9\right)$
55. Factoring the polynomial: $30a^4b + 35a^3b^2 - 15a^2b^3 = 5a^2b\left(6a^2 + 7ab - 3b^2\right) = 5a^2b(3a - b)(2a + 3b)$
57. Setting each factor equal to 0:

$$x - 5 = 0 \qquad\qquad x + 2 = 0$$
$$x = 5 \qquad\qquad\qquad x = -2$$

The solutions are –2, 5.

59. Solving the equation by factoring:

$$m^2 + 3m = 10$$
$$m^2 + 3m - 10 = 0$$
$$(m + 5)(m - 2) = 0$$
$$m = -5, 2$$

61. Solving the equation by factoring:

$$m^2 - 9m = 0$$
$$m(m - 9) = 0$$
$$m = 0, 9$$

63. Solving the equation by factoring:
$$9x^4 + 9x^3 = 10x^2$$
$$9x^4 + 9x^3 - 10x^2 = 0$$
$$x^2\left(9x^2 + 9x - 10\right) = 0$$
$$x^2(3x - 2)(3x + 5) = 0$$
$$x = 0, \frac{2}{3}, -\frac{5}{3}$$

65. Let x and $x + 1$ represent the two integers. The equation is:
$$x(x + 1) = 110$$
$$x^2 + x = 110$$
$$x^2 + x - 110 = 0$$
$$(x + 11)(x - 10) = 0$$
$$x = -11, 10$$
$$x + 1 = -10, 11$$
The two integers are either -11 and -10, or 10 and 11.

67. Let x and $20 - x$ represent the two numbers. The equation is:
$$x(20 - x) = 75$$
$$20x - x^2 = 75$$
$$0 = x^2 - 20x + 75$$
$$0 = (x - 15)(x - 5)$$
$$x = 15, 5$$
$$20 - x = 5, 15$$
The two numbers are 5 and 15.

69. Let b represent the base, and $8b$ represent the height. The equation is:
$$\frac{1}{2}(b)(8b) = 16$$
$$4b^2 = 16$$
$$4b^2 - 16 = 0$$
$$4\left(b^2 - 4\right) = 0$$
$$4(b + 2)(b - 2) = 0$$
$$b = 2 \quad \left(b = -2 \text{ is impossible}\right)$$
The base is 2 inches.

Chapters 1-6 Cumulative Review

1. Simplifying the expression: $-|-9| = -9$

3. Simplifying the expression: $20 - (-9) = 20 + 9 = 29$

5. Simplifying the expression: $\dfrac{9(-2)}{-2} = \dfrac{-18}{-2} = 9$

7. Simplifying the expression: $\dfrac{-3(4 - 7) - 5(7 - 2)}{-5 - 2 - 1} = \dfrac{-3(-3) - 5(5)}{-8} = \dfrac{9 - 25}{-8} = \dfrac{-16}{-8} = 2$

9. Simplifying the expression: $6 - 2(4a + 2) - 5 = 6 - 8a - 4 - 5 = -8a - 3$

11. Simplifying using the rules of exponents: $(9xy)^0 = 1$

13. Simplifying using the rules of exponents: $\dfrac{50x^8y^8}{25x^4y^2} + \dfrac{28x^7y^7}{14x^3y} = 2x^4y^6 + 2x^4y^6 = 4x^4y^6$

15.   Solving the equation:

$$3x = -18$$
$$\tfrac{1}{3}(3x) = \tfrac{1}{3}(-18)$$
$$x = -6$$

17.   Solving the equation:

$$-\frac{x}{3} = 7$$
$$-3\left(-\frac{x}{3}\right) = -3(7)$$
$$x = -21$$

19.   Setting each factor equal to 0:

$$
\begin{array}{lll}
4m = 0 & m - 7 = 0 & 2m - 7 = 0 \\
m = 0 & m = 7 & 2m = 7 \\
& & m = \tfrac{7}{2}
\end{array}
$$

The solutions are $0, 7,$ and $\tfrac{7}{2}$.

21.   Solving the inequality:

$$-2x > -8$$
$$-\tfrac{1}{2}(-2x) < -\tfrac{1}{2}(-8)$$
$$x < 4$$

23.   Graphing the line:

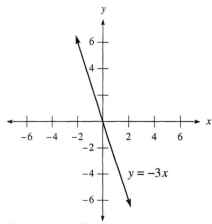

25.   Checking the point $(0,0)$: $2(0) + 3(0) = 0 \geq 6$ \quad (false)
Graphing the linear inequality:

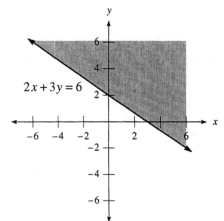

27.   Finding the slope: $m = \dfrac{-4-3}{-2-7} = \dfrac{-7}{-9} = \dfrac{7}{9}$

29.   The slope-intercept form is $y = -\tfrac{2}{5}x - \tfrac{2}{3}$.

31. Graphing both lines:

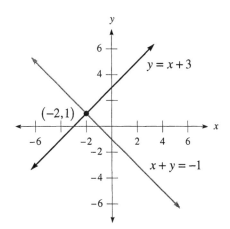

The intersection point is $(-2,1)$.

33. Multiplying the first equation by -3 and the second equation by 7:
$$-15x - 21y = 54$$
$$56x + 21y = 28$$
Adding the two equations:
$$41x = 82$$
$$x = 2$$
Substituting into the second equation:
$$8(2) + 3y = 4$$
$$16 + 3y = 4$$
$$3y = -12$$
$$y = -4$$
The solution is $(2,-4)$.

35. Multiplying the first equation by -0.04:
$$-0.04x - 0.04y = -200$$
$$0.04x + 0.06y = 270$$
Adding the two equations:
$$0.02y = 70$$
$$y = 3500$$
Substituting into the first equation:
$$x + 3500 = 5000$$
$$x = 1500$$
The solution is $(1500, 3500)$.

37. Factoring the polynomial: $n^2 - 5n - 36 = (n-9)(n+4)$

39. Factoring the polynomial: $16 - a^2 = (4+a)(4-a)$

41. Factoring the polynomial: $45x^2y - 30xy^2 + 5y^3 = 5y(9x^2 - 6xy + y^2) = 5y(3x-y)(3x-y) = 5y(3x-y)^2$

43. Factoring the polynomial: $3xy + 15x - 2y - 10 = 3x(y+5) - 2(y+5) = (y+5)(3x-2)$

45. Commutative property of addition

47. Dividing by the monomial: $\dfrac{28x^4y^4 - 14x^2y^3 + 21xy^2}{-7xy^2} = \dfrac{28x^4y^4}{-7xy^2} + \dfrac{-14x^2y^3}{-7xy^2} + \dfrac{21xy^2}{-7xy^2} = -4x^3y^2 + 2xy - 3$

49. Let x and $x + 4$ represent the length of each piece. The equation is:
$$x + x + 4 = 72$$
$$2x + 4 = 72$$
$$2x = 68$$
$$x = 34$$
$$x + 4 = 38$$
The pieces are 34 inches and 38 inches in length.

Chapter 6 Test

1. Factoring the polynomial: $5x - 10 = 5(x - 2)$
2. Factoring the polynomial: $18x^2 y - 9xy - 36xy^2 = 9xy(2x - 1 - 4y)$
3. Factoring the polynomial: $x^2 + 2ax - 3bx - 6ab = x(x + 2a) - 3b(x + 2a) = (x + 2a)(x - 3b)$
4. Factoring the polynomial: $xy + 4x - 7y - 28 = x(y + 4) - 7(y + 4) = (y + 4)(x - 7)$
5. Factoring the polynomial: $x^2 - 5x + 6 = (x - 2)(x - 3)$ **6.**   Factoring the polynomial: $x^2 - x - 6 = (x - 3)(x + 2)$
7. Factoring the polynomial: $a^2 - 16 = (a + 4)(a - 4)$ **8.**   The polynomial $x^2 + 25$ cannot be factored.
9. Factoring the polynomial: $x^4 - 81 = \left(x^2 + 9\right)\left(x^2 - 9\right) = \left(x^2 + 9\right)(x + 3)(x - 3)$
10. Factoring the polynomial: $27x^2 - 75y^2 = 3\left(9x^2 - 25y^2\right) = 3(3x + 5y)(3x - 5y)$
11. Factoring the polynomial: $x^3 + 5x^2 - 9x - 45 = x^2(x + 5) - 9(x + 5) = (x + 5)\left(x^2 - 9\right) = (x + 5)(x + 3)(x - 3)$
12. Factoring the polynomial: $x^2 - bx + 5x - 5b = x(x - b) + 5(x - b) = (x - b)(x + 5)$
13. Factoring the polynomial: $4a^2 + 22a + 10 = 2\left(2a^2 + 11a + 5\right) = 2(2a + 1)(a + 5)$
14. Factoring the polynomial: $3m^2 - 3m - 18 = 3\left(m^2 - m - 6\right) = 3(m - 3)(m + 2)$
15. Factoring the polynomial: $6y^2 + 7y - 5 = (3y + 5)(2y - 1)$
16. Factoring the polynomial: $12x^3 - 14x^2 - 10x = 2x\left(6x^2 - 7x - 5\right) = 2x(3x - 5)(2x + 1)$

17. Solving the equation by factoring:

$$x^2 + 7x + 12 = 0$$
$$(x + 4)(x + 3) = 0$$
$$x = -4, -3$$

18. Solving the equation by factoring:

$$x^2 - 4x + 4 = 0$$
$$(x - 2)^2 = 0$$
$$x - 2 = 0$$
$$x = 2$$

19. Solving the equation by factoring:

$$x^2 - 36 = 0$$
$$(x + 6)(x - 6) = 0$$
$$x = -6, 6$$

20. Solving the equation by factoring:

$$x^2 = x + 20$$
$$x^2 - x - 20 = 0$$
$$(x + 4)(x - 5) = 0$$
$$x = -4, 5$$

21. Solving the equation by factoring:

$$x^2 - 11x = -30$$
$$x^2 - 11x + 30 = 0$$
$$(x - 6)(x - 5) = 0$$
$$x = 5, 6$$

22. Solving the equation by factoring:

$$y^3 = 16y$$
$$y^3 - 16y = 0$$
$$y\left(y^2 - 16\right) = 0$$
$$y(y + 4)(y - 4) = 0$$
$$y = 0, -4, 4$$

23. Solving the equation by factoring:

$$2a^2 = a + 15$$
$$2a^2 - a - 15 = 0$$
$$(2a + 5)(a - 3) = 0$$
$$a = -\tfrac{5}{2}, 3$$

24. Solving the equation by factoring:

$$30x^3 - 20x^2 = 10x$$
$$30x^3 - 20x^2 - 10x = 0$$
$$10x\left(3x^2 - 2x - 1\right) = 0$$
$$10x(3x + 1)(x - 1) = 0$$
$$x = 0, -\tfrac{1}{3}, 1$$

25. Let x and $20 - x$ represent the numbers. The equation is:

$$x(20 - x) = 64$$
$$20x - x^2 = 64$$
$$0 = x^2 - 20x + 64$$
$$0 = (x - 16)(x - 4)$$
$$x = 4, 16$$
$$20 - x = 16, 4$$

The numbers are 4 and 16.

26. Let x and $x + 2$ represent the two integers. The equation is:

$$x(x + 2) = x + x + 2 + 7$$
$$x^2 + 2x = 2x + 9$$
$$x^2 - 9 = 0$$
$$(x + 3)(x - 3) = 0$$
$$x = -3, 3$$
$$x + 2 = -1, 5$$

The integers are either -3 and -1, or 3 and 5.

27. Let w represent the width and $3w + 5$ represent the length. The equation is:

$$w(3w + 5) = 42$$
$$3w^2 + 5w = 42$$
$$3w^2 + 5w - 42 = 0$$
$$(3w + 14)(w - 3) = 0$$
$$w = 3 \qquad \left(w = -\tfrac{14}{3} \text{ is impossible}\right)$$
$$3w + 5 = 14$$

The width is 3 feet and the length is 14 feet.

28. Let x and $2x + 2$ represent the two legs. The equation is:

$$x^2 + (2x + 2)^2 = 13^2$$
$$x^2 + 4x^2 + 8x + 4 = 169$$
$$5x^2 + 8x - 165 = 0$$
$$(5x + 33)(x - 5) = 0$$
$$x = 5 \qquad \left(x = -\tfrac{33}{5} \text{ is impossible}\right)$$
$$2x + 2 = 12$$

The two legs are 5 meters and 12 meters in length.

29. Setting $C = \$800$:

$$800 = 200 + 500x - 100x^2$$
$$100x^2 - 500x + 600 = 0$$
$$100\left(x^2 - 5x + 6\right) = 0$$
$$100(x - 2)(x - 3) = 0$$
$$x = 2, 3$$

The company can manufacture either 200 items or 300 items.

30. The revenue is given by: $R = xp = (900 - 100p)p$. Setting $R = \$1,800$:

$$1800 = (900 - 100p)p$$
$$1800 = 900p - 100p^2$$
$$100p^2 - 900p + 1800 = 0$$
$$100\left(p^2 - 9p + 18\right) = 0$$
$$100(p - 6)(p - 3) = 0$$
$$p = 3, 6$$

The manufacturer should sell the items at either $3 or $6.

Chapter 7
Rational Expressions

7.1 Reducing Rational Expressions to Lowest Terms

1. Reducing the rational expression: $\dfrac{5}{5x-10} = \dfrac{5}{5(x-2)} = \dfrac{1}{x-2}$. The variable restriction is $x \neq 2$.

3. Reducing the rational expression: $\dfrac{a-3}{a^2-9} = \dfrac{1(a-3)}{(a+3)(a-3)} = \dfrac{1}{a+3}$. The variable restriction is $a \neq -3, 3$.

5. Reducing the rational expression: $\dfrac{x+5}{x^2-25} = \dfrac{1(x+5)}{(x+5)(x-5)} = \dfrac{1}{x-5}$. The variable restriction is $x \neq -5, 5$.

7. Reducing the rational expression: $\dfrac{2x^2-8}{4} = \dfrac{2\left(x^2-4\right)}{4} = \dfrac{2(x+2)(x-2)}{4} = \dfrac{(x+2)(x-2)}{2}$

   There are no variable restrictions.

9. Reducing the rational expression: $\dfrac{2x-10}{3x-6} = \dfrac{2(x-5)}{3(x-2)}$. The variable restriction is $x \neq 2$.

11. Reducing the rational expression: $\dfrac{10a+20}{5a+10} = \dfrac{10(a+2)}{5(a+2)} = \dfrac{2}{1} = 2$

13. Reducing the rational expression: $\dfrac{5x^2-5}{4x+4} = \dfrac{5\left(x^2-1\right)}{4(x+1)} = \dfrac{5(x+1)(x-1)}{4(x+1)} = \dfrac{5(x-1)}{4}$

15. Reducing the rational expression: $\dfrac{x-3}{x^2-6x+9} = \dfrac{1(x-3)}{(x-3)^2} = \dfrac{1}{x-3}$

17. Reducing the rational expression: $\dfrac{3x+15}{3x^2+24x+45} = \dfrac{3(x+5)}{3\left(x^2+8x+15\right)} = \dfrac{3(x+5)}{3(x+5)(x+3)} = \dfrac{1}{x+3}$

19. Reducing the rational expression: $\dfrac{a^2-3a}{a^3-8a^2+15a} = \dfrac{a(a-3)}{a\left(a^2-8a+15\right)} = \dfrac{a(a-3)}{a(a-3)(a-5)} = \dfrac{1}{a-5}$

21. Reducing the rational expression: $\dfrac{3x-2}{9x^2-4} = \dfrac{1(3x-2)}{(3x+2)(3x-2)} = \dfrac{1}{3x+2}$

23. Reducing the rational expression: $\dfrac{x^2+8x+15}{x^2+5x+6} = \dfrac{(x+5)(x+3)}{(x+2)(x+3)} = \dfrac{x+5}{x+2}$

25. Reducing the rational expression: $\dfrac{2m^3 - 2m^2 - 12m}{m^2 - 5m + 6} = \dfrac{2m(m^2 - m - 6)}{(m-2)(m-3)} = \dfrac{2m(m-3)(m+2)}{(m-2)(m-3)} = \dfrac{2m(m+2)}{m-2}$

27. Reducing the rational expression: $\dfrac{x^3 + 3x^2 - 4x}{x^3 - 16x} = \dfrac{x(x^2 + 3x - 4)}{x(x^2 - 16)} = \dfrac{x(x+4)(x-1)}{x(x+4)(x-4)} = \dfrac{x-1}{x-4}$

29. Reducing the rational expression: $\dfrac{4x^3 - 10x^2 + 6x}{2x^3 + x^2 - 3x} = \dfrac{2x(2x^2 - 5x + 3)}{x(2x^2 + x - 3)} = \dfrac{2x(2x-3)(x-1)}{x(2x+3)(x-1)} = \dfrac{2(2x-3)}{2x+3}$

31. Reducing the rational expression: $\dfrac{4x^2 - 12x + 9}{4x^2 - 9} = \dfrac{(2x-3)^2}{(2x+3)(2x-3)} = \dfrac{2x-3}{2x+3}$

33. Reducing the rational expression: $\dfrac{x+3}{x^4 - 81} = \dfrac{x+3}{(x^2+9)(x^2-9)} = \dfrac{x+3}{(x^2+9)(x+3)(x-3)} = \dfrac{1}{(x^2+9)(x-3)}$

35. Reducing the rational expression: $\dfrac{3x^2 + x - 10}{x^4 - 16} = \dfrac{(3x-5)(x+2)}{(x^2+4)(x^2-4)} = \dfrac{(3x-5)(x+2)}{(x^2+4)(x+2)(x-2)} = \dfrac{3x-5}{(x^2+4)(x-2)}$

37. Reducing the rational expression: $\dfrac{42x^3 - 20x^2 - 48x}{6x^2 - 5x - 4} = \dfrac{2x(21x^2 - 10x - 24)}{(3x-4)(2x+1)} = \dfrac{2x(7x+6)(3x-4)}{(3x-4)(2x+1)} = \dfrac{2x(7x+6)}{2x+1}$

39. Reducing the rational expression: $\dfrac{xy + 3x + 2y + 6}{xy + 3x + 5y + 15} = \dfrac{x(y+3) + 2(y+3)}{x(y+3) + 5(y+3)} = \dfrac{(y+3)(x+2)}{(y+3)(x+5)} = \dfrac{x+2}{x+5}$

41. Reducing the rational expression: $\dfrac{x^2 - 3x + ax - 3a}{x^2 - 3x + bx - 3b} = \dfrac{x(x-3) + a(x-3)}{x(x-3) + b(x-3)} = \dfrac{(x-3)(x+a)}{(x-3)(x+b)} = \dfrac{x+a}{x+b}$

43. Reducing the rational expression: $\dfrac{xy + bx + ay + ab}{xy + bx + 3y + 3b} = \dfrac{x(y+b) + a(y+b)}{x(y+b) + 3(y+b)} = \dfrac{(y+b)(x+a)}{(y+b)(x+3)} = \dfrac{x+a}{x+3}$

45. Writing as a ratio: $\dfrac{8}{6} = \dfrac{4}{3}$ 47. Writing as a ratio: $\dfrac{200}{250} = \dfrac{4}{5}$

49. Writing as a ratio: $\dfrac{32}{4} = \dfrac{8}{1}$

51. Completing the table:

Checks Written x	Total Cost $2.00 + 0.15x$	Cost per Check $\dfrac{2.00 + 0.15x}{x}$
0	$2.00	undefined
5	$2.75	$0.55
10	$3.50	$0.35
15	$4.25	$0.28
20	$5.00	$0.25

53. The average speed is: $\dfrac{122 \text{ miles}}{3 \text{ hours}} \approx 40.7 \text{ miles / hour}$

55. The average speed is: $\dfrac{785 \text{ feet}}{20 \text{ minutes}} = 39.25 \text{ feet / minute}$

57. The average speed is: $\dfrac{518 \text{ feet}}{40 \text{ seconds}} = 12.95 \text{ feet / second}$

59. Her average speed on level ground is: $\dfrac{20 \text{ minutes}}{2 \text{ miles}} = 10 \text{ minutes / mile}$, or $\dfrac{2 \text{ miles}}{20 \text{ minutes}} = 0.1 \text{ miles / minute}$

      Her average speed downhill is: $\dfrac{40 \text{ minutes}}{6 \text{ miles}} = \dfrac{20}{3} \text{ minutes / mile}$, or $\dfrac{6 \text{ miles}}{40 \text{ minutes}} = \dfrac{3}{20} \text{ miles / minute}$

61. The average fuel consumption is: $\dfrac{168 \text{ miles}}{3.5 \text{ gallons}} = 48 \text{ miles / gallon}$

63. Substituting $x = 5$ and $y = 4$: $\dfrac{x^2 - y^2}{x - y} = \dfrac{5^2 - 4^2}{5 - 4} = \dfrac{25 - 16}{5 - 4} = \dfrac{9}{1} = 9$. The result is equal to $5 + 4 = 9$.

65. Completing the table:

x	$\dfrac{x-3}{3-x}$
-2	-1
-1	-1
0	-1
1	-1
2	-1

67. Completing the table:

x	$\dfrac{x-5}{x^2-25}$	$\dfrac{1}{x+5}$
0	$\frac{1}{5}$	$\frac{1}{5}$
-2	$\frac{1}{3}$	$\frac{1}{3}$
2	$\frac{1}{7}$	$\frac{1}{7}$
-5	undefined	undefined
5	undefined	$\frac{1}{10}$

Simplifying: $\dfrac{x-3}{3-x} = \dfrac{-(3-x)}{3-x} = -1$

69. Simplifying the expression: $\dfrac{27x^5}{9x^2} - \dfrac{45x^8}{15x^5} = 3x^3 - 3x^3 = 0$

71. Simplifying the expression: $\dfrac{72a^3b^7}{9ab^5} + \dfrac{64a^5b^3}{8a^3b} = 8a^2b^2 + 8a^2b^2 = 16a^2b^2$

73. Dividing by the monomial: $\dfrac{38x^7 + 42x^5 - 84x^3}{2x^3} = \dfrac{38x^7}{2x^3} + \dfrac{42x^5}{2x^3} - \dfrac{84x^3}{2x^3} = 19x^4 + 21x^2 - 42$

75. Dividing by the monomial: $\dfrac{28a^5b^5 + 36ab^4 - 44a^4b}{4ab} = \dfrac{28a^5b^5}{4ab} + \dfrac{36ab^4}{4ab} - \dfrac{44a^4b}{4ab} = 7a^4b^4 + 9b^3 - 11a^3$

7.2 Multiplication and Division of Rational Expressions

1. Simplifying the expression: $\dfrac{x+y}{3} \cdot \dfrac{6}{x+y} = \dfrac{6(x+y)}{3(x+y)} = 2$

3. Simplifying the expression: $\dfrac{2x+10}{x^2} \cdot \dfrac{x^3}{4x+20} = \dfrac{2(x+5)}{x^2} \cdot \dfrac{x^3}{4(x+5)} = \dfrac{2x^3(x+5)}{4x^2(x+5)} = \dfrac{x}{2}$

5. Simplifying the expression: $\dfrac{9}{2a-8} \div \dfrac{3}{a-4} = \dfrac{9}{2a-8} \cdot \dfrac{a-4}{3} = \dfrac{9}{2(a-4)} \cdot \dfrac{a-4}{3} = \dfrac{9(a-4)}{6(a-4)} = \dfrac{3}{2}$

7. Simplifying the expression:
$$\dfrac{x+1}{x^2-9} \div \dfrac{2x+2}{x+3} = \dfrac{x+1}{x^2-9} \cdot \dfrac{x+3}{2x+2} = \dfrac{x+1}{(x+3)(x-3)} \cdot \dfrac{x+3}{2(x+1)} = \dfrac{(x+1)(x+3)}{2(x+3)(x-3)(x+1)} = \dfrac{1}{2(x-3)}$$

9. Simplifying the expression: $\dfrac{a^2+5a}{7a} \cdot \dfrac{4a^2}{a^2+4a} = \dfrac{a(a+5)}{7a} \cdot \dfrac{4a^2}{a(a+4)} = \dfrac{4a^3(a+5)}{7a^2(a+4)} = \dfrac{4a(a+5)}{7(a+4)}$

11. Simplifying the expression:
$$\dfrac{y^2-5y+6}{2y+4} \div \dfrac{2y-6}{y+2} = \dfrac{y^2-5y+6}{2y+4} \cdot \dfrac{y+2}{2y-6} = \dfrac{(y-2)(y-3)}{2(y+2)} \cdot \dfrac{y+2}{2(y-3)} = \dfrac{(y-2)(y-3)(y+2)}{4(y+2)(y-3)} = \dfrac{y-2}{4}$$

13. Simplifying the expression:
$$\dfrac{2x-8}{x^2-4} \cdot \dfrac{x^2+6x+8}{x-4} = \dfrac{2(x-4)}{(x+2)(x-2)} \cdot \dfrac{(x+4)(x+2)}{x-4} = \dfrac{2(x-4)(x+4)(x+2)}{(x+2)(x-2)(x-4)} = \dfrac{2(x+4)}{x-2}$$

15. Simplifying the expression:
$$\dfrac{x-1}{x^2-x-6} \cdot \dfrac{x^2+5x+6}{x^2-1} = \dfrac{x-1}{(x-3)(x+2)} \cdot \dfrac{(x+2)(x+3)}{(x+1)(x-1)} = \dfrac{(x-1)(x+2)(x+3)}{(x-3)(x+2)(x+1)(x-1)} = \dfrac{x+3}{(x-3)(x+1)}$$

17. Simplifying the expression:
$$\dfrac{a^2+10a+25}{a+5} \div \dfrac{a^2-25}{a-5} = \dfrac{a^2+10a+25}{a+5} \cdot \dfrac{a-5}{a^2-25} = \dfrac{(a+5)^2}{a+5} \cdot \dfrac{a-5}{(a+5)(a-5)} = \dfrac{(a+5)^2(a-5)}{(a+5)^2(a-5)} = 1$$

19.   Simplifying the expression:

$$\frac{y^3 - 5y^2}{y^4 + 3y^3 + 2y^2} \div \frac{y^2 - 5y + 6}{y^2 - 2y - 3} = \frac{y^3 - 5y^2}{y^4 + 3y^3 + 2y^2} \cdot \frac{y^2 - 2y - 3}{y^2 - 5y + 6}$$

$$= \frac{y^2(y-5)}{y^2(y+2)(y+1)} \cdot \frac{(y-3)(y+1)}{(y-2)(y-3)}$$

$$= \frac{y^2(y-5)(y-3)(y+1)}{y^2(y+2)(y+1)(y-2)(y-3)}$$

$$= \frac{y-5}{(y+2)(y-2)}$$

21.   Simplifying the expression:

$$\frac{2x^2 + 17x + 21}{x^2 + 2x - 35} \cdot \frac{x^2 - 25}{2x^2 - 7x - 15} = \frac{(2x+3)(x+7)}{(x+7)(x-5)} \cdot \frac{(x+5)(x-5)}{(2x+3)(x-5)} = \frac{(2x+3)(x+7)(x+5)(x-5)}{(x+7)(x-5)^2(2x+3)} = \frac{x+5}{x-5}$$

23.   Simplifying the expression:

$$\frac{2x^2 + 10x + 12}{4x^2 + 24x + 32} \cdot \frac{2x^2 + 18x + 40}{x^2 + 8x + 15} = \frac{2(x^2 + 5x + 6)}{4(x^2 + 6x + 8)} \cdot \frac{2(x^2 + 9x + 20)}{x^2 + 8x + 15}$$

$$= \frac{2(x+2)(x+3)}{4(x+4)(x+2)} \cdot \frac{2(x+5)(x+4)}{(x+5)(x+3)}$$

$$= \frac{4(x+2)(x+3)(x+4)(x+5)}{4(x+2)(x+3)(x+4)(x+5)}$$

$$= 1$$

25.   Simplifying the expression:

$$\frac{2a^2 + 7a + 3}{a^2 - 16} \div \frac{4a^2 + 8a + 3}{2a^2 - 5a - 12} = \frac{2a^2 + 7a + 3}{a^2 - 16} \cdot \frac{2a^2 - 5a - 12}{4a^2 + 8a + 3}$$

$$= \frac{(2a+1)(a+3)}{(a+4)(a-4)} \cdot \frac{(2a+3)(a-4)}{(2a+1)(2a+3)}$$

$$= \frac{(2a+1)(a+3)(2a+3)(a-4)}{(a+4)(a-4)(2a+1)(2a+3)}$$

$$= \frac{a+3}{a+4}$$

27.   Simplifying the expression:

$$\frac{4y^2 - 12y + 9}{y^2 - 36} \div \frac{2y^2 - 5y + 3}{y^2 + 5y - 6} = \frac{4y^2 - 12y + 9}{y^2 - 36} \cdot \frac{y^2 + 5y - 6}{2y^2 - 5y + 3}$$

$$= \frac{(2y-3)^2}{(y+6)(y-6)} \cdot \frac{(y+6)(y-1)}{(2y-3)(y-1)}$$

$$= \frac{(2y-3)^2(y+6)(y-1)}{(y+6)(y-6)(2y-3)(y-1)}$$

$$= \frac{2y-3}{y-6}$$

29.   Simplifying the expression:

$$\frac{x^2 - 1}{6x^2 + 42x + 60} \cdot \frac{7x^2 + 17x + 6}{x+1} \cdot \frac{6x + 30}{7x^2 - 11x - 6} = \frac{(x+1)(x-1)}{6(x+5)(x+2)} \cdot \frac{(7x+3)(x+2)}{x+1} \cdot \frac{6(x+5)}{(7x+3)(x-2)}$$

$$= \frac{6(x+1)(x-1)(7x+3)(x+2)(x+5)}{6(x+5)(x+2)(x+1)(7x+3)(x-2)}$$

$$= \frac{x-1}{x-2}$$

31. Simplifying the expression:

$$\frac{18x^3 + 21x^2 - 60x}{21x^2 - 25x - 4} \cdot \frac{28x^2 - 17x - 3}{16x^3 + 28x^2 - 30x} = \frac{3x\left(6x^2 + 7x - 20\right)}{21x^2 - 25x - 4} \cdot \frac{28x^2 - 17x - 3}{2x\left(8x^2 + 14x - 15\right)}$$

$$= \frac{3x(3x - 4)(2x + 5)}{(7x + 1)(3x - 4)} \cdot \frac{(7x + 1)(4x - 3)}{2x(4x - 3)(2x + 5)}$$

$$= \frac{3x(3x - 4)(2x + 5)(7x + 1)(4x - 3)}{2x(7x + 1)(3x - 4)(4x - 3)(2x + 5)}$$

$$= \frac{3}{2}$$

33. Simplifying the expression: $\left(x^2 - 9\right)\left(\dfrac{2}{x + 3}\right) = \dfrac{(x + 3)(x - 3)}{1} \cdot \dfrac{2}{x + 3} = \dfrac{2(x + 3)(x - 3)}{x + 3} = 2(x - 3)$

35. Simplifying the expression: $a(a + 5)(a - 5)\left(\dfrac{2}{a^2 - 25}\right) = \dfrac{a(a + 5)(a - 5)}{1} \cdot \dfrac{2}{(a + 5)(a - 5)} = \dfrac{2a(a + 5)(a - 5)}{(a + 5)(a - 5)} = 2a$

37. Simplifying the expression: $\left(x^2 - x - 6\right)\left(\dfrac{x + 1}{x - 3}\right) = \dfrac{(x - 3)(x + 2)}{1} \cdot \dfrac{x + 1}{x - 3} = \dfrac{(x - 3)(x + 2)(x + 1)}{x - 3} = (x + 2)(x + 1)$

39. Simplifying the expression: $\left(x^2 - 4x - 5\right)\left(\dfrac{-2x}{x + 1}\right) = \dfrac{(x - 5)(x + 1)}{1} \cdot \dfrac{-2x}{x + 1} = \dfrac{-2x(x - 5)(x + 1)}{x + 1} = -2x(x - 5)$

41. Simplifying the expression:

$$\frac{x^2 - 9}{x^2 - 3x} \cdot \frac{2x + 10}{xy + 5x + 3y + 15} = \frac{(x + 3)(x - 3)}{x(x - 3)} \cdot \frac{2(x + 5)}{x(y + 5) + 3(y + 5)} = \frac{2(x + 3)(x - 3)(x + 5)}{x(x - 3)(y + 5)(x + 3)} = \frac{2(x + 5)}{x(y + 5)}$$

43. Simplifying the expression:

$$\frac{2x^2 + 4x}{x^2 - y^2} \cdot \frac{x^2 + 3x + xy + 3y}{x^2 + 5x + 6} = \frac{2x(x + 2)}{(x + y)(x - y)} \cdot \frac{x(x + 3) + y(x + 3)}{(x + 2)(x + 3)} = \frac{2x(x + 2)(x + 3)(x + y)}{(x + y)(x - y)(x + 2)(x + 3)} = \frac{2x}{x - y}$$

45. Simplifying the expression:

$$\frac{x^3 - 3x^2 + 4x - 12}{x^4 - 16} \cdot \frac{3x^2 + 5x - 2}{3x^2 - 10x + 3} = \frac{x^2(x - 3) + 4(x - 3)}{\left(x^2 + 4\right)\left(x^2 - 4\right)} \cdot \frac{(3x - 1)(x + 2)}{(3x - 1)(x - 3)}$$

$$= \frac{(x - 3)\left(x^2 + 4\right)}{\left(x^2 + 4\right)(x + 2)(x - 2)} \cdot \frac{(3x - 1)(x + 2)}{(3x - 1)(x - 3)}$$

$$= \frac{(x - 3)\left(x^2 + 4\right)(3x - 1)(x + 2)}{\left(x^2 + 4\right)(x + 2)(x - 2)(3x - 1)(x - 3)}$$

$$= \frac{1}{x - 2}$$

47. Simplifying the expression: $\left(1 - \frac{1}{2}\right)\left(1 - \frac{1}{3}\right)\left(1 - \frac{1}{4}\right)\left(1 - \frac{1}{5}\right) = \left(\frac{2}{2} - \frac{1}{2}\right)\left(\frac{3}{3} - \frac{1}{3}\right)\left(\frac{4}{4} - \frac{1}{4}\right)\left(\frac{5}{5} - \frac{1}{5}\right) = \frac{1}{2} \cdot \frac{2}{3} \cdot \frac{3}{4} \cdot \frac{4}{5} = \frac{1}{5}$

49. Simplifying the expression: $\left(1 - \frac{1}{2}\right)\left(1 - \frac{1}{3}\right)\left(1 - \frac{1}{4}\right) \bullet \bullet \bullet \left(1 - \frac{1}{99}\right)\left(1 - \frac{1}{100}\right) = \frac{1}{2} \cdot \frac{2}{3} \cdot \frac{3}{4} \bullet \bullet \bullet \frac{98}{99} \cdot \frac{99}{100} = \frac{1}{100}$

51. Since 5,280 feet = 1 mile, the height is: $\dfrac{14,494 \text{ feet}}{5,280 \text{ feet / mile}} \approx 2.7$ miles

53. Converting to miles per hour: $\dfrac{1088 \text{ feet}}{1 \text{ second}} \cdot \dfrac{1 \text{ mile}}{5280 \text{ feet}} \cdot \dfrac{60 \text{ seconds}}{1 \text{ minute}} \cdot \dfrac{60 \text{ minutes}}{1 \text{ hour}} \approx 742$ miles / hour

55. Converting to miles per hour: $\dfrac{785 \text{ feet}}{20 \text{ minutes}} \cdot \dfrac{60 \text{ minutes}}{1 \text{ hour}} \cdot \dfrac{1 \text{ mile}}{5280 \text{ feet}} \approx 0.45$ miles / hour

57. Converting to miles per hour: $\dfrac{518 \text{ feet}}{40 \text{ seconds}} \cdot \dfrac{60 \text{ seconds}}{1 \text{ minute}} \cdot \dfrac{60 \text{ minutes}}{1 \text{ hour}} \cdot \dfrac{1 \text{ mile}}{5280 \text{ feet}} \approx 8.8$ miles / hour

59. Her average speed on level ground is: $\dfrac{2 \text{ miles}}{1 / 3 \text{ hour}} = 6$ miles / hour

Her average speed downhill is: $\dfrac{6 \text{ miles}}{2 / 3 \text{ hour}} = 9$ miles / hour

61.   Adding the fractions: $\frac{1}{2} + \frac{5}{2} = \frac{6}{2} = 3$

63.   Adding the fractions: $2 + \frac{3}{4} = \frac{2 \cdot 4}{1 \cdot 4} + \frac{3}{4} = \frac{8}{4} + \frac{3}{4} = \frac{11}{4}$

65.   Adding the fractions: $\frac{1}{10} + \frac{3}{14} = \frac{1 \cdot 7}{10 \cdot 7} + \frac{3 \cdot 5}{14 \cdot 5} = \frac{7}{70} + \frac{15}{70} = \frac{22}{70} = \frac{11}{35}$

67.   Simplifying the expression: $\frac{10x^4}{2x^2} + \frac{12x^6}{3x^4} = 5x^2 + 4x^2 = 9x^2$

69.   Simplifying the expression: $\frac{12a^2b^5}{3ab^3} + \frac{14a^4b^7}{7a^3b^5} = 4ab^2 + 2ab^2 = 6ab^2$

7.3 Addition and Subtraction of Rational Expressions

1.   Combining the fractions: $\frac{3}{x} + \frac{4}{x} = \frac{7}{x}$

3.   Combining the fractions: $\frac{9}{a} - \frac{5}{a} = \frac{4}{a}$

5.   Combining the fractions: $\frac{1}{x+1} + \frac{x}{x+1} = \frac{1+x}{x+1} = 1$

7.   Combining the fractions: $\frac{y^2}{y-1} - \frac{1}{y-1} = \frac{y^2-1}{y-1} = \frac{(y+1)(y-1)}{y-1} = y+1$

9.   Combining the fractions: $\frac{x^2}{x+2} + \frac{4x+4}{x+2} = \frac{x^2+4x+4}{x+2} = \frac{(x+2)^2}{x+2} = x+2$

11.   Combining the fractions: $\frac{x^2}{x-2} - \frac{4x-4}{x-2} = \frac{x^2-4x+4}{x-2} = \frac{(x-2)^2}{x-2} = x-2$

13.   Combining the fractions: $\frac{x+2}{x+6} - \frac{x-4}{x+6} = \frac{x+2-x+4}{x+6} = \frac{6}{x+6}$

15.   Combining the fractions: $\frac{y}{2} - \frac{2}{y} = \frac{y \cdot y}{2 \cdot y} - \frac{2 \cdot 2}{y \cdot 2} = \frac{y^2}{2y} - \frac{4}{2y} = \frac{y^2-4}{2y} = \frac{(y+2)(y-2)}{2y}$

17.   Combining the fractions: $\frac{1}{2} + \frac{a}{3} = \frac{1 \cdot 3}{2 \cdot 3} + \frac{a \cdot 2}{3 \cdot 2} = \frac{3}{6} + \frac{2a}{6} = \frac{2a+3}{6}$

19.   Combining the fractions: $\frac{x}{x+1} + \frac{3}{4} = \frac{x \cdot 4}{(x+1) \cdot 4} + \frac{3 \cdot (x+1)}{4 \cdot (x+1)} = \frac{4x}{4(x+1)} + \frac{3x+3}{4(x+1)} = \frac{4x+3x+3}{4(x+1)} = \frac{7x+3}{4(x+1)}$

21.   Combining the fractions:
$$\frac{x+1}{x-2} - \frac{4x+7}{5x-10} = \frac{(x+1) \cdot 5}{(x-2) \cdot 5} - \frac{4x+7}{5(x-2)} = \frac{5x+5}{5(x-2)} - \frac{4x+7}{5(x-2)} = \frac{5x+5-4x-7}{5(x-2)} = \frac{x-2}{5(x-2)} = \frac{1}{5}$$

23.   Combining the fractions:
$$\frac{4x-2}{3x+12} - \frac{x-2}{x+4} = \frac{4x-2}{3(x+4)} - \frac{(x-2) \cdot 3}{(x+4) \cdot 3} = \frac{4x-2}{3(x+4)} - \frac{3x-6}{3(x+4)} = \frac{4x-2-3x+6}{3(x+4)} = \frac{x+4}{3(x+4)} = \frac{1}{3}$$

25.   Combining the fractions:
$$\frac{6}{x(x-2)} + \frac{3}{x} = \frac{6}{x(x-2)} + \frac{3 \cdot (x-2)}{x \cdot (x-2)} = \frac{6}{x(x-2)} + \frac{3x-6}{x(x-2)} = \frac{6+3x-6}{x(x-2)} = \frac{3x}{x(x-2)} = \frac{3}{x-2}$$

27.   Combining the fractions:
$$\frac{4}{a} - \frac{12}{a^2+3a} = \frac{4 \cdot (a+3)}{a \cdot (a+3)} - \frac{12}{a(a+3)} = \frac{4a+12}{a(a+3)} - \frac{12}{a(a+3)} = \frac{4a+12-12}{a(a+3)} = \frac{4a}{a(a+3)} = \frac{4}{a+3}$$

29.   Combining the fractions:

$$\frac{2}{x+5} - \frac{10}{x^2 - 25} = \frac{2 \cdot (x-5)}{(x+5) \cdot (x-5)} - \frac{10}{(x+5)(x-5)}$$

$$= \frac{2x-10}{(x+5)(x-5)} - \frac{10}{(x+5)(x-5)}$$

$$= \frac{2x-10-10}{(x+5)(x-5)}$$

$$= \frac{2x-20}{(x+5)(x-5)}$$

$$= \frac{2(x-10)}{(x+5)(x-5)}$$

31.   Combining the fractions:

$$\frac{x-4}{x-3} + \frac{6}{x^2-9} = \frac{(x-4) \cdot (x+3)}{(x-3) \cdot (x+3)} + \frac{6}{(x+3)(x-3)}$$

$$= \frac{x^2-x-12}{(x+3)(x-3)} + \frac{6}{(x+3)(x-3)}$$

$$= \frac{x^2-x-12+6}{(x+3)(x-3)}$$

$$= \frac{x^2-x-6}{(x+3)(x-3)}$$

$$= \frac{(x-3)(x+2)}{(x+3)(x-3)}$$

$$= \frac{x+2}{x+3}$$

33.   Combining the fractions:

$$\frac{a-4}{a-3} + \frac{5}{a^2-a-6} = \frac{(a-4) \cdot (a+2)}{(a-3) \cdot (a+2)} + \frac{5}{(a-3)(a+2)}$$

$$= \frac{a^2-2a-8}{(a-3)(a+2)} + \frac{5}{(a-3)(a+2)}$$

$$= \frac{a^2-2a-8+5}{(a-3)(a+2)}$$

$$= \frac{a^2-2a-3}{(a-3)(a+2)}$$

$$= \frac{(a-3)(a+1)}{(a-3)(a+2)}$$

$$= \frac{a+1}{a+2}$$

35.   Combining the fractions:

$$\frac{8}{x^2-16} - \frac{7}{x^2-x-12} = \frac{8}{(x+4)(x-4)} - \frac{7}{(x-4)(x+3)}$$

$$= \frac{8(x+3)}{(x+4)(x-4)(x+3)} - \frac{7(x+4)}{(x+4)(x-4)(x+3)}$$

$$= \frac{8x+24}{(x+4)(x-4)(x+3)} - \frac{7x+28}{(x+4)(x-4)(x+3)}$$

$$= \frac{8x+24-7x-28}{(x+4)(x-4)(x+3)}$$

$$= \frac{x-4}{(x+4)(x-4)(x+3)}$$

$$= \frac{1}{(x+4)(x+3)}$$

37. Combining the fractions:

$$\frac{4y}{y^2+6y+5}-\frac{3y}{y^2+5y+4}=\frac{4y}{(y+5)(y+1)}-\frac{3y}{(y+4)(y+1)}$$

$$=\frac{4y(y+4)}{(y+5)(y+1)(y+4)}-\frac{3y(y+5)}{(y+5)(y+1)(y+4)}$$

$$=\frac{4y^2+16y}{(y+5)(y+1)(y+4)}-\frac{3y^2+15y}{(y+5)(y+1)(y+4)}$$

$$=\frac{4y^2+16y-3y^2-15y}{(y+5)(y+1)(y+4)}$$

$$=\frac{y^2+y}{(y+5)(y+1)(y+4)}$$

$$=\frac{y(y+1)}{(y+5)(y+1)(y+4)}$$

$$=\frac{y}{(y+5)(y+4)}$$

39. Combining the fractions:

$$\frac{4x+1}{x^2+5x+4}-\frac{x+3}{x^2+4x+3}=\frac{4x+1}{(x+4)(x+1)}-\frac{x+3}{(x+3)(x+1)}$$

$$=\frac{(4x+1)(x+3)}{(x+4)(x+1)(x+3)}-\frac{(x+3)(x+4)}{(x+4)(x+1)(x+3)}$$

$$=\frac{4x^2+13x+3}{(x+4)(x+1)(x+3)}-\frac{x^2+7x+12}{(x+4)(x+1)(x+3)}$$

$$=\frac{4x^2+13x+3-x^2-7x-12}{(x+4)(x+1)(x+3)}$$

$$=\frac{3x^2+6x-9}{(x+4)(x+1)(x+3)}$$

$$=\frac{3(x+3)(x-1)}{(x+4)(x+1)(x+3)}$$

$$=\frac{3(x-1)}{(x+4)(x+1)}$$

41. Combining the fractions:

$$\frac{1}{x}+\frac{x}{3x+9}-\frac{3}{x^2+3x}=\frac{1}{x}+\frac{x}{3(x+3)}-\frac{3}{x(x+3)}$$

$$=\frac{1\cdot3(x+3)}{x\cdot3(x+3)}+\frac{x\cdot x}{3(x+3)\cdot x}-\frac{3\cdot3}{x(x+3)\cdot3}$$

$$=\frac{3x+9}{3x(x+3)}+\frac{x^2}{3x(x+3)}-\frac{9}{3x(x+3)}$$

$$=\frac{3x+9+x^2-9}{3x(x+3)}$$

$$=\frac{x^2+3x}{3x(x+3)}$$

$$=\frac{x(x+3)}{3x(x+3)}$$

$$=\frac{1}{3}$$

43. Completing the table:

Number x	Reciprocal $\dfrac{1}{x}$	Sum $1+\dfrac{1}{x}$	Sum $\dfrac{x+1}{x}$
1	1	2	$\dfrac{2}{2}$
2	$\dfrac{1}{2}$	$\dfrac{3}{2}$	$\dfrac{3}{2}$
3	$\dfrac{1}{3}$	$\dfrac{4}{3}$	$\dfrac{4}{3}$
4	$\dfrac{1}{4}$	$\dfrac{5}{4}$	$\dfrac{5}{4}$

45. Completing the table:

x	$x+\dfrac{4}{x}$	$\dfrac{x^2+4}{x}$	$x+4$
1	5	5	5
2	4	4	6
3	$\dfrac{13}{3}$	$\dfrac{13}{3}$	7
4	5	5	8

47. Combining the fractions: $1+\dfrac{1}{x+2}=\dfrac{1\bullet(x+2)}{1\bullet(x+2)}+\dfrac{1}{x+2}=\dfrac{x+2}{x+2}+\dfrac{1}{x+2}=\dfrac{x+2+1}{x+2}=\dfrac{x+3}{x+2}$

49. Combining the fractions: $1-\dfrac{1}{x+3}=\dfrac{1\bullet(x+3)}{1\bullet(x+3)}-\dfrac{1}{x+3}=\dfrac{x+3}{x+3}-\dfrac{1}{x+3}=\dfrac{x+3-1}{x+3}=\dfrac{x+2}{x+3}$

51. The expression is: $x+2\left(\dfrac{1}{x}\right)=x+\dfrac{2}{x}=\dfrac{x\bullet x}{1\bullet x}+\dfrac{2}{x}=\dfrac{x^2}{x}+\dfrac{2}{x}=\dfrac{x^2+2}{x}$

53. Represent the two numbers as x and $2x$. The expression is: $\dfrac{1}{x}+\dfrac{1}{2x}=\dfrac{1\bullet2}{x\bullet2}+\dfrac{1}{2x}=\dfrac{2}{2x}+\dfrac{1}{2x}=\dfrac{3}{2x}$

55. Solving the equation:
$$2x+3(x-3)=6$$
$$2x+3x-9=6$$
$$5x-9=6$$
$$5x=15$$
$$x=3$$

57. Solving the equation:
$$x-3(x+3)=x-3$$
$$x-3x-9=x-3$$
$$-2x-9=x-3$$
$$-3x-9=-3$$
$$-3x=6$$
$$x=-2$$

59. Solving the equation:
$$7-2(3x+1)=4x+3$$
$$7-6x-2=4x+3$$
$$-6x+5=4x+3$$
$$-10x+5=3$$
$$-10x=-2$$
$$x=\tfrac{1}{5}$$

61. Solving the equation:
$$x^2+5x+6=0$$
$$(x+2)(x+3)=0$$
$$x=-2,-3$$

63. Solving the quadratic equation:
$$x^2-x=6$$
$$x^2-x-6=0$$
$$(x-3)(x+2)=0$$
$$x=-2,3$$

65. Solving the quadratic equation:
$$x^2-5x=0$$
$$x(x-5)=0$$
$$x=0,5$$

7.4 Equations Involving Rational Expressions

1. Multiplying both sides of the equation by 6:
$$6\left(\dfrac{x}{3}+\dfrac{1}{2}\right)=6\left(-\dfrac{1}{2}\right)$$
$$2x+3=-3$$
$$2x=-6$$
$$x=-3$$
Since $x=-3$ checks in the original equation, the solution is $x=-3$.

3. Multiplying both sides of the equation by $5a$:
$$5a\left(\dfrac{4}{a}\right)=5a\left(\dfrac{1}{5}\right)$$
$$20=a$$
Since $a=20$ checks in the original equation, the solution is $a=20$.

5.   Multiplying both sides of the equation by x:
$$x\left(\frac{3}{x}+1\right) = x\left(\frac{2}{x}\right)$$
$$3+x=2$$
$$x=-1$$
Since $x=-1$ checks in the original equation, the solution is $x=-1$.

7.   Multiplying both sides of the equation by $5a$:
$$5a\left(\frac{3}{a}-\frac{2}{a}\right) = 5a\left(\frac{1}{5}\right)$$
$$15-10=a$$
$$a=5$$
Since $a=5$ checks in the original equation, the solution is $a=5$.

9.   Multiplying both sides of the equation by $2x$:
$$2x\left(\frac{3}{x}+2\right) = 2x\left(\frac{1}{2}\right)$$
$$6+4x=x$$
$$6=-3x$$
$$x=-2$$
Since $x=-2$ checks in the original equation, the solution is $x=-2$.

11.   Multiplying both sides of the equation by $4y$:
$$4y\left(\frac{1}{y}-\frac{1}{2}\right) = 4y\left(-\frac{1}{4}\right)$$
$$4-2y=-y$$
$$4=y$$
Since $y=4$ checks in the original equation, the solution is $y=4$.

13.   Multiplying both sides of the equation by x^2:
$$x^2\left(1-\frac{8}{x}\right) = x^2\left(-\frac{15}{x^2}\right)$$
$$x^2-8x=-15$$
$$x^2-8x+15=0$$
$$(x-3)(x-5)=0$$
$$x=3,5$$
Both $x=3$ and $x=5$ check in the original equation.

15.   Multiplying both sides of the equation by $2x$:
$$2x\left(\frac{x}{2}-\frac{4}{x}\right) = 2x\left(-\frac{7}{2}\right)$$
$$x^2-8=-7x$$
$$x^2+7x-8=0$$
$$(x+8)(x-1)=0$$
$$x=-8,1$$
Both $x=-8$ and $x=1$ check in the original equation.

17.   Multiplying both sides of the equation by 6:
$$6\left(\frac{x-3}{2}+\frac{2x}{3}\right) = 6\left(\frac{5}{6}\right)$$
$$3(x-3)+2(2x)=5$$
$$3x-9+4x=5$$
$$7x-9=5$$
$$7x=14$$
$$x=2$$
Since $x=2$ checks in the original equation, the solution is $x=2$.

19.   Multiplying both sides of the equation by 12:
$$12\left(\frac{x+1}{3}+\frac{x-3}{4}\right) = 12\left(\frac{1}{6}\right)$$
$$4(x+1)+3(x-3)=2$$
$$4x+4+3x-9=2$$
$$7x-5=2$$
$$7x=7$$
$$x=1$$
Since $x=1$ checks in the original equation, the solution is $x=1$.

21. Multiplying both sides of the equation by $5(x+2)$:

$$5(x+2) \cdot \frac{6}{x+2} = 5(x+2) \cdot \frac{3}{5}$$
$$30 = 3x+6$$
$$24 = 3x$$
$$x = 8$$

Since $x = 8$ checks in the original equation, the solution is $x = 8$.

23. Multiplying both sides of the equation by $(y-2)(y-3)$:

$$(y-2)(y-3) \cdot \frac{3}{y-2} = (y-2)(y-3) \cdot \frac{2}{y-3}$$
$$3(y-3) = 2(y-2)$$
$$3y-9 = 2y-4$$
$$y = 5$$

Since $y = 5$ checks in the original equation, the solution is $y = 5$.

25. Multiplying both sides of the equation by $3(x-2)$:

$$3(x-2)\left(\frac{x}{x-2} + \frac{2}{3}\right) = 3(x-2)\left(\frac{2}{x-2}\right)$$
$$3x + 2(x-2) = 6$$
$$3x + 2x - 4 = 6$$
$$5x - 4 = 6$$
$$5x = 10$$
$$x = 2$$

Since $x = 2$ does not check in the original equation, there is no solution.

27. Multiplying both sides of the equation by $2(x-2)$:

$$2(x-2)\left(\frac{x}{x-2} + \frac{3}{2}\right) = 2(x-2) \cdot \frac{9}{2(x-2)}$$
$$2x + 3(x-2) = 9$$
$$2x + 3x - 6 = 9$$
$$5x - 6 = 9$$
$$5x = 15$$
$$x = 3$$

Since $x = 3$ checks in the original equation, the solution is $x = 3$.

29. Multiplying both sides of the equation by $x^2 + 5x + 6 = (x+2)(x+3)$:

$$(x+2)(x+3)\left(\frac{5}{x+2} + \frac{1}{x+3}\right) = (x+2)(x+3) \cdot \frac{-1}{(x+2)(x+3)}$$
$$5(x+3) + 1(x+2) = -1$$
$$5x + 15 + x + 2 = -1$$
$$6x + 17 = -1$$
$$6x = -18$$
$$x = -3$$

Since $x = -3$ does not check in the original equation, there is no solution.

31. Multiplying both sides of the equation by $x^2 - 4 = (x+2)(x-2)$:

$$(x+2)(x-2)\left(\frac{8}{x^2-4} + \frac{3}{x+2}\right) = (x+2)(x-2) \cdot \frac{1}{x-2}$$
$$8 + 3(x-2) = 1(x+2)$$
$$8 + 3x - 6 = x+2$$
$$3x + 2 = x+2$$
$$2x = 0$$
$$x = 0$$

Since $x = 0$ checks in the original equation, the solution is $x = 0$.

152   Chapter 7   Rational Expressions

33.   Multiplying both sides of the equation by $2(a-3)$:

$$2(a-3)\left(\frac{a}{2}+\frac{3}{a-3}\right)=2(a-3)\cdot\frac{a}{a-3}$$
$$a(a-3)+6=2a$$
$$a^2-3a+6=2a$$
$$a^2-5a+6=0$$
$$(a-2)(a-3)=0$$
$$a=2,3$$

Since $a=3$ does not check in the original equation, the solution is $a=2$.

35.   Since $y^2-4=(y+2)(y-2)$ and $y^2+2y=y(y+2)$, the LCD is $y(y+2)(y-2)$. Multiplying by the LCD:

$$y(y+2)(y-2)\cdot\frac{6}{(y+2)(y-2)}=y(y+2)(y-2)\cdot\frac{4}{y(y+2)}$$
$$6y=4(y-2)$$
$$6y=4y-8$$
$$2y=-8$$
$$y=-4$$

Since $y=-4$ checks in the original equation, the solution is $y=-4$.

37.   Since $a^2-9=(a+3)(a-3)$ and $a^2+a-12=(a+4)(a-3)$, the LCD is $(a+3)(a-3)(a+4)$. Multiplying by the LCD:

$$(a+3)(a-3)(a+4)\cdot\frac{2}{(a+3)(a-3)}=(a+3)(a-3)(a+4)\cdot\frac{3}{(a+4)(a-3)}$$
$$2(a+4)=3(a+3)$$
$$2a+8=3a+9$$
$$-a+8=9$$
$$-a=1$$
$$a=-1$$

Since $a=-1$ checks in the original equation, the solution is $a=-1$.

39.   Multiplying both sides of the equation by $x^2-4x-5=(x-5)(x+1)$:

$$(x-5)(x+1)\left(\frac{3x}{x-5}-\frac{2x}{x+1}\right)=(x-5)(x+1)\cdot\frac{-42}{(x-5)(x+1)}$$
$$3x(x+1)-2x(x-5)=-42$$
$$3x^2+3x-2x^2+10x=-42$$
$$x^2+13x+42=0$$
$$(x+7)(x+6)=0$$
$$x=-7,-6$$

Both $x=-7$ and $x=-6$ check in the original equation.

41.   Multiplying both sides of the equation by $x^2+5x+6=(x+2)(x+3)$:

$$(x+2)(x+3)\left(\frac{2x}{x+2}\right)=(x+2)(x+3)\left(\frac{x}{x+3}-\frac{3}{x^2+5x+6}\right)$$
$$2x(x+3)=x(x+2)-3$$
$$2x^2+6x=x^2+2x-3$$
$$x^2+4x+3=0$$
$$(x+3)(x+1)=0$$
$$x=-3,-1$$

Since $x=-3$ does not check in the original equation, the solution is $x=-1$.

43. Solving the equation:
$$x + \frac{4}{x} = 5$$
$$x\left(x + \frac{4}{x}\right) = x(5)$$
$$x^2 + 4 = 5x$$
$$x^2 - 5x + 4 = 0$$
$$(x-1)(x-4) = 0$$
$$x = 1, 4$$
The solution is consistent with the table.

45. Let x represent the number. The equation is:
$$2(x-3) - 5 = 3$$
$$2x - 6 - 5 = 3$$
$$2x - 11 = 3$$
$$2x = 14$$
$$x = 7$$
The number is 7.

47. Let w represent the width, and $2w + 5$ represent the length. Using the perimeter formula:
$$2w + 2(2w + 5) = 34$$
$$2w + 4w + 10 = 34$$
$$6w + 10 = 34$$
$$6w = 24$$
$$w = 4$$
$$2w + 5 = 13$$
The length is 13 inches and the width is 4 inches.

49. Let x and $x + 2$ represent the two integers. The equation is:
$$x(x+2) = 48$$
$$x^2 + 2x = 48$$
$$x^2 + 2x - 48 = 0$$
$$(x+8)(x-6) = 0$$
$$x = -8, 6$$
$$x + 2 = -6, 8$$
The two integers are either –8 and –6, or 6 and 8.

51. Let x and $x + 2$ represent the two legs. The equation is:
$$x^2 + (x+2)^2 = 10^2$$
$$x^2 + x^2 + 4x + 4 = 100$$
$$2x^2 + 4x - 96 = 0$$
$$x^2 + 2x - 48 = 0$$
$$(x+8)(x-6) = 0$$
$$x = 6 \quad (x = -8 \text{ is impossible})$$
$$x + 2 = 8$$
The legs are 6 inches and 8 inches.

7.5 Applications

1. Let x and $3x$ represent the two numbers. The equation is:
$$\frac{1}{x} + \frac{1}{3x} = \frac{16}{3}$$
$$3x\left(\frac{1}{x} + \frac{1}{3x}\right) = 3x\left(\frac{16}{3}\right)$$
$$3 + 1 = 16x$$
$$16x = 4$$
$$x = \frac{1}{4}$$
$$3x = \frac{3}{4}$$
The numbers are $\frac{1}{4}$ and $\frac{3}{4}$.

3.   Let x represent the number. The equation is:

$$x + \frac{1}{x} = \frac{13}{6}$$

$$6x\left(x + \frac{1}{x}\right) = 6x\left(\frac{13}{6}\right)$$

$$6x^2 + 6 = 13x$$

$$6x^2 - 13x + 6 = 0$$

$$(3x - 2)(2x - 3) = 0$$

$$x = \frac{2}{3}, \frac{3}{2}$$

The number is either $\frac{2}{3}$ or $\frac{3}{2}$.

5.   Let x represent the number. The equation is:

$$\frac{7 + x}{9 + x} = \frac{5}{7}$$

$$7(9 + x) \cdot \frac{7 + x}{9 + x} = 7(9 + x) \cdot \frac{5}{7}$$

$$7(7 + x) = 5(9 + x)$$

$$49 + 7x = 45 + 5x$$

$$49 + 2x = 45$$

$$2x = -4$$

$$x = -2$$

The number is –2.

7.   Let x and $x + 2$ represent the two integers. The equation is:

$$\frac{1}{x} + \frac{1}{x + 2} = \frac{5}{12}$$

$$12x(x + 2)\left(\frac{1}{x} + \frac{1}{x + 2}\right) = 12x(x + 2)\left(\frac{5}{12}\right)$$

$$12(x + 2) + 12x = 5x(x + 2)$$

$$12x + 24 + 12x = 5x^2 + 10x$$

$$0 = 5x^2 - 14x - 24$$

$$(5x + 6)(x - 4) = 0$$

$$x = 4 \quad \left(x = -\frac{6}{5} \text{ is impossible}\right)$$

$$x + 2 = 6$$

The integers are 4 and 6.

9.   Let x represent the rate of the boat in still water:

	d	r	t
Upstream	26	$x - 3$	$\dfrac{26}{x - 3}$
Downstream	38	$x + 3$	$\dfrac{38}{x + 3}$

The equation is:

$$\frac{26}{x - 3} = \frac{38}{x + 3}$$

$$(x + 3)(x - 3) \cdot \frac{26}{x - 3} = (x + 3)(x - 3) \cdot \frac{38}{x + 3}$$

$$26(x + 3) = 38(x - 3)$$

$$26x + 78 = 38x - 114$$

$$-12x + 78 = -114$$

$$-12x = -192$$

$$x = 16$$

The speed of the boat in still water is 16 mph.

11.   Let x represent the plane speed in still air:

	d	r	t
Against Wind	140	$x - 20$	$\dfrac{140}{x - 20}$
With Wind	160	$x + 20$	$\dfrac{160}{x + 20}$

The equation is:

$$\frac{140}{x - 20} = \frac{160}{x + 20}$$

$$(x + 20)(x - 20) \cdot \frac{140}{x - 20} = (x + 20)(x - 20) \cdot \frac{160}{x + 20}$$

$$140(x + 20) = 160(x - 20)$$

$$140x + 2800 = 160x - 3200$$

$$-20x + 2800 = -3200$$

$$-20x = -6000$$

$$x = 300$$

The plane speed in still air is 300 mph.

15.   Let x represent her rate downhill:

	d	r	t
Level Ground	2	$x - 3$	$\dfrac{2}{x - 3}$
Downhill	6	x	$\dfrac{6}{x}$

The equation is:

$$\frac{2}{x - 3} + \frac{6}{x} = 1$$

$$x(x - 3)\left(\frac{2}{x - 3} + \frac{6}{x}\right) = x(x - 3) \cdot 1$$

$$2x + 6(x - 3) = x(x - 3)$$

$$2x + 6x - 18 = x^2 - 3x$$

$$8x - 18 = x^2 - 3x$$

$$0 = x^2 - 11x + 18$$

$$0 = (x - 2)(x - 9)$$

$$x = 9 \quad (x = 2 \text{ is impossible})$$

Tina runs 9 mph on the downhill part of the course.

17.   Let x represent her rate on level ground:

	d	r	t
Level Ground	4	x	$\dfrac{4}{x}$
Downhill	5	$x + 2$	$\dfrac{5}{x + 2}$

The equation is:

$$\frac{4}{x} + \frac{5}{x + 2} = 1$$

$$x(x + 2)\left(\frac{4}{x} + \frac{5}{x + 2}\right) = x(x + 2) \cdot 1$$

$$4(x + 2) + 5x = x(x + 2)$$

$$4x + 8 + 5x = x^2 + 2x$$

$$9x + 8 = x^2 + 2x$$

$$0 = x^2 - 7x - 8$$

$$0 = (x - 8)(x + 1)$$

$$x = 8 \quad (x = -1 \text{ is impossible})$$

Jerri jogs 8 mph on level ground.

13.   Let x and $x + 20$ represent the rates of each plane:

	d	r	t
Plane 1	285	$x + 20$	$\dfrac{285}{x + 20}$
Plane 2	255	x	$\dfrac{255}{x}$

The equation is:

$$\frac{285}{x + 20} = \frac{255}{x}$$

$$x(x + 20) \cdot \frac{285}{x + 20} = x(x + 20) \cdot \frac{255}{x}$$

$$285x = 255(x + 20)$$

$$285x = 255x + 5100$$

$$30x = 5100$$

$$x = 170$$

$$x + 20 = 190$$

The plane speeds are 170 mph and 190 mph.

19.   Let t represent the time to fill the pool with both pipes left open. The equation is:
$$\frac{1}{12} - \frac{1}{15} = \frac{1}{t}$$
$$60t\left(\frac{1}{12} - \frac{1}{15}\right) = 60t \cdot \frac{1}{t}$$
$$5t - 4t = 60$$
$$t = 60$$
It will take 60 hours to fill the pool with both pipes left open.

21.   Let t represent the time to fill the bathtub with both faucets open. The equation is:
$$\frac{1}{10} + \frac{1}{12} = \frac{1}{t}$$
$$60t\left(\frac{1}{10} + \frac{1}{12}\right) = 60t \cdot \frac{1}{t}$$
$$6t + 5t = 60$$
$$11t = 60$$
$$t = \frac{60}{11} = 5\frac{5}{11}$$
It will take $5\frac{5}{11}$ minutes to fill the tub with both faucets open.

23.   Let t represent the time to fill the sink with both the faucet and the drain left open. The equation is:
$$\frac{1}{3} - \frac{1}{4} = \frac{1}{t}$$
$$12t\left(\frac{1}{3} - \frac{1}{4}\right) = 12t \cdot \frac{1}{t}$$
$$4t - 3t = 12$$
$$t = 12$$
It will take 12 minutes for the sink to overflow with both the faucet and drain left open.

25.   Sketching the line graph:

27.   Sketching the graph:

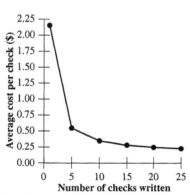

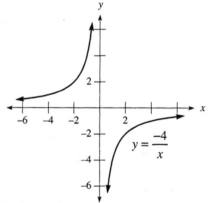

29.   Sketching the graph:

31.   Sketching the graph:

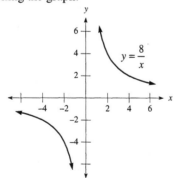

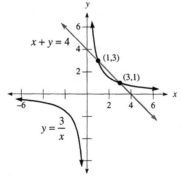

The intersection points are (1,3) and (3,1).

33. Factoring the polynomial: $15a^3b^3 - 20a^2b - 35ab^2 = 5ab\left(3a^2b^2 - 4a - 7b\right)$

35. Factoring the polynomial: $x^2 - 4x - 12 = (x-6)(x+2)$

37. Factoring the polynomial: $x^4 - 16 = \left(x^2 + 4\right)\left(x^2 - 4\right) = \left(x^2 + 4\right)(x+2)(x-2)$

39. Factoring the polynomial: $5x^3 - 25x^2 - 30x = 5x\left(x^2 - 5x - 6\right) = 5x(x-6)(x+1)$

41. Solving the equation by factoring:

$$x^2 - 6x = 0$$
$$x(x-6) = 0$$
$$x = 0, 6$$

43. Solving the equation by factoring:

$$x(x+2) = 80$$
$$x^2 + 2x = 80$$
$$x^2 + 2x - 80 = 0$$
$$(x+10)(x-8) = 0$$
$$x = -10, 8$$

45. Let x and $x+3$ represent the two integers. The equation is:

$$x^2 + (x+3)^2 = 15^2$$
$$x^2 + x^2 + 6x + 9 = 225$$
$$2x^2 + 6x - 216 = 0$$
$$x^2 + 3x - 108 = 0$$
$$(x+12)(x-9) = 0$$
$$x = 9 \quad \left(x = -12 \text{ is impossible}\right)$$
$$x + 3 = 12$$

The two legs are 9 inches and 12 inches.

7.6 Complex Fractions

1. Simplifying the complex fraction: $\dfrac{\frac{3}{4}}{\frac{7}{8}} = \dfrac{\frac{3}{4} \bullet 8}{\frac{7}{8} \bullet 8} = \dfrac{6}{1} = 6$

3. Simplifying the complex fraction: $\dfrac{\frac{2}{3}}{4} = \dfrac{\frac{2}{3} \bullet 3}{4 \bullet 3} = \dfrac{2}{12} = \dfrac{1}{6}$

5. Simplifying the complex fraction: $\dfrac{\frac{x^2}{y}}{\frac{x}{y^3}} = \dfrac{\frac{x^2}{y} \bullet y^3}{\frac{x}{y^3} \bullet y^3} = \dfrac{x^2 y^2}{x} = xy^2$

7. Simplifying the complex fraction: $\dfrac{\frac{4x^3}{y^6}}{\frac{8x^2}{y^7}} = \dfrac{\frac{4x^3}{y^6} \bullet y^7}{\frac{8x^2}{y^7} \bullet y^7} = \dfrac{4x^3 y}{8x^2} = \dfrac{xy}{2}$

9. Simplifying the complex fraction: $\dfrac{y + \frac{1}{x}}{x + \frac{1}{y}} = \dfrac{\left(y + \frac{1}{x}\right) \bullet xy}{\left(x + \frac{1}{y}\right) \bullet xy} = \dfrac{xy^2 + y}{x^2 y + x} = \dfrac{y(xy+1)}{x(xy+1)} = \dfrac{y}{x}$

11. Simplifying the complex fraction: $\dfrac{1 + \frac{1}{a}}{1 - \frac{1}{a}} = \dfrac{\left(1 + \frac{1}{a}\right) \bullet a}{\left(1 - \frac{1}{a}\right) \bullet a} = \dfrac{a+1}{a-1}$

13. Simplifying the complex fraction: $\dfrac{\frac{x+1}{x^2-9}}{\frac{2}{x+3}} = \dfrac{\frac{x+1}{(x+3)(x-3)} \bullet (x+3)(x-3)}{\frac{2}{x+3} \bullet (x+3)(x-3)} = \dfrac{x+1}{2(x-3)}$

15.   Simplifying the complex fraction: $\dfrac{\dfrac{1}{a+2}}{\dfrac{1}{a^2-a-6}} = \dfrac{\dfrac{1}{a+2}\bullet(a-3)(a+2)}{\dfrac{1}{(a-3)(a+2)}\bullet(a-3)(a+2)} = \dfrac{a-3}{1} = a-3$

17.   Simplifying the complex fraction: $\dfrac{1-\dfrac{9}{y^2}}{1-\dfrac{1}{y}-\dfrac{6}{y^2}} = \dfrac{\left(1-\dfrac{9}{y^2}\right)\bullet y^2}{\left(1-\dfrac{1}{y}-\dfrac{6}{y^2}\right)\bullet y^2} = \dfrac{y^2-9}{y^2-y-6} = \dfrac{(y+3)(y-3)}{(y+2)(y-3)} = \dfrac{y+3}{y+2}$

19.   Simplifying the complex fraction: $\dfrac{\dfrac{1}{y}+\dfrac{1}{x}}{\dfrac{1}{xy}} = \dfrac{\left(\dfrac{1}{y}+\dfrac{1}{x}\right)\bullet xy}{\left(\dfrac{1}{xy}\right)\bullet xy} = \dfrac{x+y}{1} = x+y$

21.   Simplifying the complex fraction: $\dfrac{1-\dfrac{1}{a^2}}{1-\dfrac{1}{a}} = \dfrac{\left(1-\dfrac{1}{a^2}\right)\bullet a^2}{\left(1-\dfrac{1}{a}\right)\bullet a^2} = \dfrac{a^2-1}{a^2-a} = \dfrac{(a+1)(a-1)}{a(a-1)} = \dfrac{a+1}{a}$

23.   Simplifying the complex fraction: $\dfrac{\dfrac{1}{10x}-\dfrac{y}{10x^2}}{\dfrac{1}{10}-\dfrac{y}{10x}} = \dfrac{\left(\dfrac{1}{10x}-\dfrac{y}{10x^2}\right)\bullet 10x^2}{\left(\dfrac{1}{10}-\dfrac{y}{10x}\right)\bullet 10x^2} = \dfrac{x-y}{x^2-xy} = \dfrac{1(x-y)}{x(x-y)} = \dfrac{1}{x}$

25.   Simplifying the complex fraction: $\dfrac{\dfrac{1}{a+1}+2}{\dfrac{1}{a+1}+3} = \dfrac{\left(\dfrac{1}{a+1}+2\right)\bullet(a+1)}{\left(\dfrac{1}{a+1}+3\right)\bullet(a+1)} = \dfrac{1+2(a+1)}{1+3(a+1)} = \dfrac{1+2a+2}{1+3a+3} = \dfrac{2a+3}{3a+4}$

27.   Simplifying each parenthesis first:
$$1-\dfrac{1}{x} = \dfrac{x}{x}-\dfrac{1}{x} = \dfrac{x-1}{x}$$
$$1-\dfrac{1}{x+1} = \dfrac{x+1}{x+1}-\dfrac{1}{x+1} = \dfrac{x}{x+1}$$
$$1-\dfrac{1}{x+2} = \dfrac{x+2}{x+2}-\dfrac{1}{x+2} = \dfrac{x+1}{x+2}$$
Now performing the multiplication: $\left(1-\dfrac{1}{x}\right)\left(1-\dfrac{1}{x+1}\right)\left(1-\dfrac{1}{x+2}\right) = \dfrac{x-1}{x}\bullet\dfrac{x}{x+1}\bullet\dfrac{x+1}{x+2} = \dfrac{x-1}{x+2}$

29.   Simplifying each parenthesis first:
$$1+\dfrac{1}{x+3} = \dfrac{x+3}{x+3}+\dfrac{1}{x+3} = \dfrac{x+4}{x+3}$$
$$1+\dfrac{1}{x+2} = \dfrac{x+2}{x+2}+\dfrac{1}{x+2} = \dfrac{x+3}{x+2}$$
$$1+\dfrac{1}{x+1} = \dfrac{x+1}{x+1}+\dfrac{1}{x+1} = \dfrac{x+2}{x+1}$$
Now performing the multiplication: $\left(1+\dfrac{1}{x+3}\right)\left(1+\dfrac{1}{x+2}\right)\left(1+\dfrac{1}{x+1}\right) = \dfrac{x+4}{x+3}\bullet\dfrac{x+3}{x+2}\bullet\dfrac{x+2}{x+1} = \dfrac{x+4}{x+1}$

31. Simplifying each term in the sequence:

$$2 + \frac{1}{2+1} = 2 + \frac{1}{3} = \frac{6}{3} + \frac{1}{3} = \frac{7}{3}$$

$$2 + \frac{1}{2 + \frac{1}{2+1}} = 2 + \frac{1}{\frac{7}{3}} = 2 + \frac{3}{7} = \frac{14}{7} + \frac{3}{7} = \frac{17}{7}$$

$$2 + \frac{2}{2 + \frac{1}{2 + \frac{1}{2+1}}} = 2 + \frac{1}{\frac{17}{7}} = 2 + \frac{7}{17} = \frac{34}{17} + \frac{7}{17} = \frac{41}{17}$$

33. Completing the table:

Number x	Reciprocal $\frac{1}{x}$	Quotient $\frac{x}{1/x}$	Square x^2
1	1	1	1
2	$\frac{1}{2}$	4	4
3	$\frac{1}{3}$	9	9
4	$\frac{1}{4}$	16	16

35. Completing the table:

Number x	Reciprocal $\frac{1}{x}$	Sum $1+\frac{1}{x}$	Quotient $\frac{1+\frac{1}{x}}{\frac{1}{x}}$
1	1	2	2
2	$\frac{1}{2}$	$\frac{3}{2}$	3
3	$\frac{1}{3}$	$\frac{4}{3}$	4
4	$\frac{1}{4}$	$\frac{5}{4}$	5

37. Solving the inequality:
$$2x+3<5$$
$$2x+3-3<5-3$$
$$2x<2$$
$$\tfrac{1}{2}(2x)<\tfrac{1}{2}(2)$$
$$x<1$$

39. Solving the inequality:
$$-3x \le 21$$
$$-\tfrac{1}{3}(-3x) \ge -\tfrac{1}{3}(21)$$
$$x \ge -7$$

41. Solving the inequality:
$$-2x+8>-4$$
$$-2x+8-8>-4-8$$
$$-2x>-12$$
$$-\tfrac{1}{2}(-2x)<-\tfrac{1}{2}(-12)$$
$$x<6$$

43. Solving the inequality:
$$4-2(x+1)\ge-2$$
$$4-2x-2\ge-2$$
$$-2x+2\ge-2$$
$$-2x+2-2\ge-2-2$$
$$-2x\ge-4$$
$$-\tfrac{1}{2}(-2x)\le-\tfrac{1}{2}(-4)$$
$$x\le2$$

7.7 Proportions

1. Solving the proportion:
$$\frac{x}{2}=\frac{6}{12}$$
$$12x=12$$
$$x=1$$

3. Solving the proportion:
$$\frac{2}{5}=\frac{4}{x}$$
$$2x=20$$
$$x=10$$

5. Solving the proportion:
$$\frac{10}{20}=\frac{20}{x}$$
$$10x=400$$
$$x=40$$

7. Solving the proportion:
$$\frac{a}{3}=\frac{5}{12}$$
$$12a=15$$
$$a=\tfrac{15}{12}=\tfrac{5}{4}$$

9.   Solving the proportion:

$$\frac{2}{x} = \frac{6}{7}$$
$$6x = 14$$
$$x = \frac{14}{6} = \frac{7}{3}$$

11.   Solving the proportion:

$$\frac{x+1}{3} = \frac{4}{x}$$
$$x^2 + x = 12$$
$$x^2 + x - 12 = 0$$
$$(x+4)(x-3) = 0$$
$$x = -4, 3$$

13.   Solving the proportion:

$$\frac{x}{2} = \frac{8}{x}$$
$$x^2 = 16$$
$$x^2 - 16 = 0$$
$$(x+4)(x-4) = 0$$
$$x = -4, 4$$

15.   Solving the proportion:

$$\frac{4}{a+2} = \frac{a}{2}$$
$$a^2 + 2a = 8$$
$$a^2 + 2a - 8 = 0$$
$$(a+4)(a-2) = 0$$
$$a = -4, 2$$

17.   Solving the proportion:

$$\frac{1}{x} = \frac{x-5}{6}$$
$$x^2 - 5x = 6$$
$$x^2 - 5x - 6 = 0$$
$$(x-6)(x+1) = 0$$
$$x = -1, 6$$

19.   Comparing hits to games, the proportion is:

$$\frac{6}{18} = \frac{x}{45}$$
$$18x = 270$$
$$x = 15$$

He will get 15 hits in 45 games.

21.   Comparing ml alcohol to ml water, the proportion is:

$$\frac{12}{16} = \frac{x}{28}$$
$$16x = 336$$
$$x = 21$$

The solution will have 21 ml of alcohol.

23.   Comparing grams of fat to total grams, the proportion is:

$$\frac{13}{100} = \frac{x}{350}$$
$$100x = 4550$$
$$x = 45.5$$

There are 45.5 grams of fat in 350 grams of ice cream.

25.   Comparing inches on the map to actual miles, the proportion is:

$$\frac{3.5}{100} = \frac{x}{420}$$
$$100x = 1470$$
$$x = 14.7$$

They are 14.7 inches apart on the map.

27.   Comparing miles to hours, the proportion is:

$$\frac{245}{5} = \frac{x}{7}$$
$$5x = 1715$$
$$x = 343$$

He will travel 343 miles.

29.   Reducing the fraction: $\dfrac{x^2 - x - 6}{x^2 - 9} = \dfrac{(x-3)(x+2)}{(x+3)(x-3)} = \dfrac{x+2}{x+3}$

31.   Multiplying the fractions:

$$\frac{x^2 - 25}{x+4} \cdot \frac{2x+8}{x^2 - 9x + 20} = \frac{(x+5)(x-5)}{x+4} \cdot \frac{2(x+4)}{(x-5)(x-4)} = \frac{2(x+5)(x-5)(x+4)}{(x+4)(x-5)(x-4)} = \frac{2(x+5)}{x-4}$$

33.   Adding the fractions: $\dfrac{x}{x^2 - 16} + \dfrac{4}{x^2 - 16} = \dfrac{x+4}{x^2 - 16} = \dfrac{1(x+4)}{(x+4)(x-4)} = \dfrac{1}{x-4}$

7.8 Variation

1. The variation equation is $y = Kx$. Finding K:
$$10 = K \cdot 5$$
$$K = 2$$
So $y = 2x$. Substituting $x = 4$: $y = 2 \cdot 4 = 8$

3. The variation equation is $y = Kx$. Finding K:
$$39 = K \cdot 3$$
$$K = 13$$
So $y = 13x$. Substituting $x = 10$: $y = 13 \cdot 10 = 130$

5. The variation equation is $y = Kx$. Finding K:
$$-24 = K \cdot 4$$
$$K = -6$$
So $y = -6x$. Substituting $y = -30$:
$$-6x = -30$$
$$x = 5$$

7. The variation equation is $y = Kx$. Finding K:
$$-7 = K \cdot (-1)$$
$$K = 7$$
So $y = 7x$. Substituting $y = -21$:
$$7x = -21$$
$$x = -3$$

9. The variation equation is $y = Kx^2$. Finding K:
$$75 = K \cdot 5^2$$
$$75 = 25K$$
$$K = 3$$
So $y = 3x^2$. Substituting $x = 1$: $y = 3 \cdot 1^2 = 3 \cdot 1 = 3$

11. The variation equation is $y = Kx^2$. Finding K:
$$48 = K \cdot 4^2$$
$$48 = 16K$$
$$K = 3$$
So $y = 3x^2$. Substituting $x = 9$: $y = 3 \cdot 9^2 = 3 \cdot 81 = 243$

13. The variation equation is $y = \dfrac{K}{x}$. Finding K:
$$5 = \frac{K}{2}$$
$$K = 10$$
So $y = \dfrac{10}{x}$. Substituting $x = 5$: $y = \dfrac{10}{5} = 2$

15. The variation equation is $y = \dfrac{K}{x}$. Finding K:
$$2 = \frac{K}{1}$$
$$K = 2$$
So $y = \dfrac{2}{x}$. Substituting $x = 4$: $y = \dfrac{2}{4} = \tfrac{1}{2}$

17. The variation equation is $y = \dfrac{K}{x}$. Finding K:
$$5 = \frac{K}{3}$$
$$K = 15$$
So $y = \dfrac{15}{x}$. Substituting $y = 15$:
$$\frac{15}{x} = 15$$
$$15 = 15x$$
$$x = 1$$

19. The variation equation is $y = \dfrac{K}{x}$. Finding K:
$$10 = \frac{K}{10}$$
$$K = 100$$
So $y = \dfrac{100}{x}$. Substituting $y = 100$:
$$\frac{100}{x} = 20$$
$$100 = 20x$$
$$x = 5$$

21. The variation equation is $y = \dfrac{K}{x^2}$. Finding K:
$$4 = \frac{K}{5^2}$$
$$4 = \frac{K}{25}$$
$$K = 100$$
So $y = \dfrac{100}{x^2}$. Substituting $x = 2$: $y = \dfrac{100}{2^2} = \dfrac{100}{4} = 25$

23. The variation equation is $y = \dfrac{K}{x^2}$. Finding K:
$$4 = \frac{K}{3^2}$$
$$4 = \frac{K}{9}$$
$$K = 36$$
So $y = \dfrac{36}{x^2}$. Substituting $x = 2$: $y = \dfrac{36}{2^2} = \dfrac{36}{4} = 9$

25. The variation equation is $t = Kd$. Finding K:
$$42 = K \cdot 2$$
$$K = 21$$
So $t = 21d$. Substituting $d = 4$: $t = 21 \cdot 4 = 84$ pounds

27. The variation equation is $P = KI^2$. Finding K:
$$30 = K \cdot 2^2$$
$$30 = 4K$$
$$K = \tfrac{15}{2}$$
So $P = \tfrac{15}{2}I^2$. Substituting $I = 7$: $P = \tfrac{15}{2} \cdot 7^2 = \tfrac{15}{2} \cdot 49 = 367.5$

29.   The variation equation is $M = Kh$. Finding K:
$$157 = K \cdot 20$$
$$K = 7.85$$
So $M = 7.85h$. Substituting $h = 30$: $M = 7.85 \cdot 30 = \$235.50$

31.   The variation equation is $F = \dfrac{K}{d^2}$. Finding K:

$$150 = \frac{K}{4000^2}$$
$$150 = \frac{K}{1.6 \times 10^7}$$
$$K = 2.4 \times 10^9$$

So $F = \dfrac{2.4 \times 10^9}{d^2}$. Substituting $d = 5000$: $F = \dfrac{2.4 \times 10^9}{(5000)^2} = \dfrac{2.4 \times 10^9}{2.5 \times 10^7} = 96$ pounds

33.   The variation equation is $I = \dfrac{K}{R}$. Finding K:

$$30 = \frac{K}{2}$$
$$K = 60$$

So $I = \dfrac{60}{R}$. Substituting $R = 5$: $I = \dfrac{60}{5} = 12$ amps

35.   Adding the two equations:
$$5x = 10$$
$$x = 2$$
Substituting into the first equation:
$$2(2) + y = 3$$
$$4 + y = 3$$
$$y = -1$$
The solution is $(2, -1)$.

37.   Multiplying the second equation by –4:
$$4x - 5y = 1$$
$$-4x + 8y = 8$$
Adding the two equations:
$$3y = 9$$
$$y = 3$$
Substituting into the second equation:
$$x - 2(3) = -2$$
$$x - 6 = -2$$
$$x = 4$$
The solution is $(4, 3)$.

39.   Substituting into the first equation:
$$5x + 2(3x - 2) = 7$$
$$5x + 6x - 4 = 7$$
$$11x - 4 = 7$$
$$11x = 11$$
$$x = 1$$
Substituting into the second equation: $y = 3(1) - 2 = 3 - 2 = 1$. The solution is $(1, 1)$.

41.   Substituting into the first equation:
$$2(2y + 1) - 3y = 4$$
$$4y + 2 - 3y = 4$$
$$y + 2 = 4$$
$$y = 2$$
Substituting into the second equation: $x = 2(2) + 1 = 4 + 1 = 5$. The solution is $(5, 2)$.

Chapter 7 Review

1. Reducing the rational expression: $\dfrac{7}{14x-28}=\dfrac{7}{14(x-2)}=\dfrac{1}{2(x-2)}$. The variable restriction is $x\neq 2$.

3. Reducing the rational expression: $\dfrac{8x-4}{4x+12}=\dfrac{4(2x-1)}{4(x+3)}=\dfrac{2x-1}{x+3}$. The variable restriction is $x\neq -3$.

5. Reducing the rational expression: $\dfrac{3x^3+16x^2-12x}{2x^3+9x^2-18x}=\dfrac{x\left(3x^2+16x-12\right)}{x\left(2x^2+9x-18\right)}=\dfrac{x(3x-2)(x+6)}{x(2x-3)(x+6)}=\dfrac{3x-2}{2x-3}$

The variable restriction is $x\neq -6,\frac{3}{2}$.

7. Reducing the rational expression: $\dfrac{x^2+5x-14}{x+7}=\dfrac{(x+7)(x-2)}{x+7}=x-2$. The variable restriction is $x\neq -7$.

9. Reducing the rational expression: $\dfrac{xy+bx+ay+ab}{xy+5x+ay+5a}=\dfrac{x(y+b)+a(y+b)}{x(y+5)+a(y+5)}=\dfrac{(y+b)(x+a)}{(y+5)(x+a)}=\dfrac{y+b}{y+5}$

The variable restriction is $y\neq -5, x\neq -a$.

11. Performing the operations:
$$\dfrac{x^2+8x+16}{x^2+x-12}\div\dfrac{x^2-16}{x^2-x-6}=\dfrac{x^2+8x+16}{x^2+x-12}\cdot\dfrac{x^2-x-6}{x^2-16}$$
$$=\dfrac{(x+4)^2}{(x+4)(x-3)}\cdot\dfrac{(x+2)(x-3)}{(x+4)(x-4)}$$
$$=\dfrac{(x+4)^2(x+2)(x-3)}{(x+4)^2(x-3)(x-4)}$$
$$=\dfrac{x+2}{x-4}$$

13. Performing the operations:
$$\dfrac{3x^2-2x-1}{x^2+6x+8}\div\dfrac{3x^2+13x+4}{x^2+8x+16}=\dfrac{3x^2-2x-1}{x^2+6x+8}\cdot\dfrac{x^2+8x+16}{3x^2+13x+4}$$
$$=\dfrac{(3x+1)(x-1)}{(x+4)(x+2)}\cdot\dfrac{(x+4)^2}{(3x+1)(x+4)}$$
$$=\dfrac{(x+4)^2(3x+1)(x-1)}{(x+4)^2(x+2)(3x+1)}$$
$$=\dfrac{x-1}{x+2}$$

15. Performing the operations: $\dfrac{x^2}{x-9}-\dfrac{18x-81}{x-9}=\dfrac{x^2-18x+81}{x-9}=\dfrac{(x-9)^2}{x-9}=x-9$

17. Performing the operations: $\dfrac{x}{x+9}+\dfrac{5}{x}=\dfrac{x\bullet x}{(x+9)\bullet x}+\dfrac{5\bullet(x+9)}{x\bullet(x+9)}=\dfrac{x^2}{x(x+9)}+\dfrac{5x+45}{x(x+9)}=\dfrac{x^2+5x+45}{x(x+9)}$

19. Performing the operations:
$$\dfrac{3}{x^2-36}-\dfrac{2}{x^2-4x-12}=\dfrac{3}{(x+6)(x-6)}-\dfrac{2}{(x-6)(x+2)}$$
$$=\dfrac{3(x+2)}{(x+6)(x-6)(x+2)}-\dfrac{2(x+6)}{(x+6)(x-6)(x+2)}$$
$$=\dfrac{3x+6}{(x+6)(x-6)(x+2)}-\dfrac{2x+12}{(x+6)(x-6)(x+2)}$$
$$=\dfrac{3x+6-2x-12}{(x+6)(x-6)(x+2)}$$
$$=\dfrac{x-6}{(x+6)(x-6)(x+2)}$$
$$=\dfrac{1}{(x+6)(x+2)}$$

21. Multiplying both sides of the equation by $2x$:

$$2x\left(\frac{3}{x}+\frac{1}{2}\right)=2x\left(\frac{5}{x}\right)$$
$$6+x=10$$
$$x=4$$

Since $x=4$ checks in the original equation, the solution is $x=4$.

23. Multiplying both sides of the equation by x^2:

$$x^2\left(1-\frac{7}{x}\right)=x^2\left(\frac{-6}{x^2}\right)$$
$$x^2-7x=-6$$
$$x^2-7x+6=0$$
$$(x-6)(x-1)=0$$
$$x=1,6$$

Both $x=1$ and $x=6$ check in the original equation.

25. Since $y^2-16=(y+4)(y-4)$ and $y^2+4y=y(y+4)$, multiply each side of the equation by $y(y+4)(y-4)$:

$$y(y+4)(y-4)\cdot\frac{2}{(y+4)(y-4)}=y(y+4)(y-4)\cdot\frac{10}{y(y+4)}$$
$$2y=10(y-4)$$
$$2y=10y-40$$
$$-8y=-40$$
$$y=5$$

Since $y=5$ checks in the original equation, the solution is $y=5$.

27. Let x represent the speed of the boat in still water. Completing the table:

	d	r	t
Upstream	48	$x-3$	$\dfrac{48}{x-3}$
Downstream	72	$x+3$	$\dfrac{72}{x+3}$

The equation is:

$$\frac{48}{x-3}=\frac{72}{x+3}$$
$$(x+3)(x-3)\cdot\frac{48}{x-3}=(x+3)(x-3)\cdot\frac{72}{x+3}$$
$$48(x+3)=72(x-3)$$
$$48x+144=72x-216$$
$$-24x+144=-216$$
$$-24x=-360$$
$$x=15$$

The speed of the boat in still water is 15 mph.

29. Simplifying the complex fraction: $\dfrac{\dfrac{x+4}{x^2-16}}{\dfrac{2}{x-4}}=\dfrac{\dfrac{x+4}{(x+4)(x-4)}}{\dfrac{2}{x-4}}=\dfrac{\dfrac{1}{x-4}\cdot(x-4)}{\dfrac{2}{x-4}\cdot(x-4)}=\dfrac{1}{2}$

31. Simplifying the complex fraction: $\dfrac{\dfrac{1}{a-2}+4}{\dfrac{1}{a-2}+1}=\dfrac{\left(\dfrac{1}{a-2}+4\right)(a-2)}{\left(\dfrac{1}{a-2}+1\right)(a-2)}=\dfrac{1+4(a-2)}{1+1(a-2)}=\dfrac{1+4a-8}{1+a-2}=\dfrac{4a-7}{a-1}$

33. Writing as a fraction: $\dfrac{40\text{ seconds}}{3\text{ minutes}}=\dfrac{40\text{ seconds}}{180\text{ seconds}}=\dfrac{2}{9}$

35. Solving the proportion:
$$\frac{a}{3} = \frac{12}{a}$$
$$a^2 = 36$$
$$a^2 - 36 = 0$$
$$(a+6)(a-6) = 0$$
$$a = -6, 6$$

37. The variation equation is $y = Kx$. Finding K:
$$-20 = K \cdot 4$$
$$K = -5$$
So $y = -5x$. Substituting $x = 7$: $y = -5 \cdot 7 = -35$

Chapters 1-7 Cumulative Review

1. Simplifying the expression: $8 - 11 = 8 + (-11) = -3$

3. Simplifying the expression: $\dfrac{-48}{12} = -4$

5. Simplifying the expression: $5x - 4 - 9x = 5x - 9x - 4 = -4x - 4$

7. Simplifying the expression: $9^{-2} = \dfrac{1}{9^2} = \frac{1}{81}$

9. Simplifying the expression: $4^1 + 9^0 + (-7)^0 = 4 + 1 + 1 = 6$

11. Simplifying the expression:
$$\left(4a^3 - 10a^2 + 6\right) - \left(6a^3 + 5a - 7\right) = 4a^3 - 10a^2 + 6 - 6a^3 - 5a + 7 = -2a^3 - 10a^2 - 5a + 13$$

13. Simplifying the expression: $\dfrac{x^2}{x-7} - \dfrac{14x - 49}{x - 7} = \dfrac{x^2 - 14x + 49}{x - 7} = \dfrac{(x-7)^2}{x-7} = x - 7$

15. Simplifying the expression: $\dfrac{\dfrac{x-2}{x^2 + 6x + 8}}{\dfrac{4}{x+4}} = \dfrac{\dfrac{x-2}{(x+4)(x+2)} \bullet (x+4)(x+2)}{\dfrac{4}{x+4} \bullet (x+4)(x+2)} = \dfrac{x-2}{4(x+2)}$

17. Solving the equation:
$$x - \tfrac{3}{4} = \tfrac{5}{6}$$
$$x = \tfrac{3}{4} + \tfrac{5}{6}$$
$$x = \tfrac{9}{12} + \tfrac{10}{12}$$
$$x = \tfrac{19}{12}$$

19. Solving the equation:
$$98r^2 - 18 = 0$$
$$2\left(49r^2 - 9\right) = 0$$
$$2(7r + 3)(7r - 3) = 0$$
$$r = -\tfrac{3}{7}, \tfrac{3}{7}$$

21. Multiplying each side of the equation by $3x$:
$$3x\left(\frac{5}{x} - \tfrac{1}{3}\right) = 3x\left(\frac{3}{x}\right)$$
$$15 - x = 9$$
$$-x = -6$$
$$x = 6$$
Since $x = 6$ checks in the original equation, the solution is $x = 6$.

23. Multiplying each side of the equation by $3(x - 3)$:
$$3(x-3) \bullet \frac{x}{3} = 3(x-3) \bullet \frac{6}{x-3}$$
$$x(x-3) = 18$$
$$x^2 - 3x = 18$$
$$x^2 - 3x - 18 = 0$$
$$(x-6)(x+3) = 0$$
$$x = 6, -3$$
Both $x = -3$ and $x = 6$ check in the original equation.

25.   Multiplying the first equation by –5 and the second equation by 3:
$$-45x - 70y = 20$$
$$45x - 24y = 27$$
Adding the two equations:
$$-94y = 47$$
$$y = -\tfrac{1}{2}$$
Substituting into the first equation:
$$9x + 14\left(-\tfrac{1}{2}\right) = -4$$
$$9x - 7 = -4$$
$$9x = 3$$
$$x = \tfrac{1}{3}$$
The solution is $\left(\tfrac{1}{3}, -\tfrac{1}{2}\right)$.

27.   To clear each equation of fractions, multiply the first equation by 6 and the second equation by 12:
$$6\left(\tfrac{1}{2}x + \tfrac{1}{3}y\right) = 6(-1) \qquad\qquad 12\left(\tfrac{1}{3}x\right) = 12\left(\tfrac{1}{4}y + 5\right)$$
$$3x + 2y = -6 \qquad\qquad\qquad\qquad 4x = 3y + 60$$
$$\qquad\qquad\qquad\qquad\qquad\qquad 4x - 3y = 60$$

The system of equations is:
$$3x + 2y = -6$$
$$4x - 3y = 60$$
Multiplying the first equation by 3 and the second equation by 2:
$$9x + 6y = -18$$
$$8x - 6y = 120$$
Adding the two equations:
$$17x = 102$$
$$x = 6$$
Substituting into $3x + 2y = -6$:
$$3(6) + 2y = -6$$
$$18 + 2y = -6$$
$$2y = -24$$
$$y = -12$$
The solution is $(6, -12)$.

29.   Graphing the inequality:

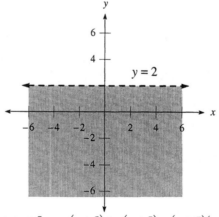

31.   Factoring the polynomial: $xy + 5x + ay + 5a = x(y+5) + a(y+5) = (y+5)(x+a)$

33.   Factoring the polynomial: $20y^2 - 27y + 9 = (5y-3)(4y-3)$

35.   Factoring the polynomial: $16x^2 + 72xy + 81y^2 = (4x+9y)(4x+9y) = (4x+9y)^2$

37. Checking each ordered pair:

$(3,-1)$: $2(3)-5=6-5=1 \neq -1$

$(1,-3)$: $2(1)-5=2-5=-3$

$(-2,9)$: $2(-2)-5=-4-5=-9 \neq 9$

The ordered pair $(1,-3)$ is a solution to the equation.

39. Solving for y:

$$3x-y=4$$
$$-y=-3x+4$$
$$y=3x-4$$

The slope is 3 and the y-intercept is -4.

41. Using the point-slope formula:

$$y-(-1)=-\tfrac{2}{5}(x-(-2))$$
$$y+1=-\tfrac{2}{5}(x+2)$$
$$y+1=-\tfrac{2}{5}x-\tfrac{4}{5}$$
$$y=-\tfrac{2}{5}x-\tfrac{9}{5}$$

43. associative property of addition

45. Simplifying the expression: $\dfrac{6x-12}{6x+12} \cdot \dfrac{3x+3}{12x-24} = \dfrac{6(x-2)}{6(x+2)} \cdot \dfrac{3(x+1)}{12(x-2)} = \dfrac{18(x-2)(x+1)}{72(x+2)(x-2)} = \dfrac{x+1}{4(x+2)}$

47. Simplifying the expression: $\dfrac{2xy+10x+3y+15}{3xy+15x+2y+10} = \dfrac{2x(y+5)+3(y+5)}{3x(y+5)+2(y+5)} = \dfrac{(y+5)(2x+3)}{(y+5)(3x+2)} = \dfrac{2x+3}{3x+2}$

49. Let x and y represent the two numbers. The system of equations is:

$$x+y=40$$
$$x-y=18$$

Adding the two equations:

$$2x=58$$
$$x=29$$

Substituting into the first equation:

$$29+y=40$$
$$y=11$$

The two numbers are 29 and 11.

Chapter 7 Test

1. Reducing the rational expression: $\dfrac{x^2-16}{x^2-8x+16} = \dfrac{(x+4)(x-4)}{(x-4)^2} = \dfrac{x+4}{x-4}$

2. Reducing the rational expression: $\dfrac{10a+20}{5a^2+20a+20} = \dfrac{10(a+2)}{5(a^2+4a+4)} = \dfrac{10(a+2)}{5(a+2)^2} = \dfrac{2}{a+2}$

3. Reducing the rational expression: $\dfrac{xy+7x+5y+35}{x^2+ax+5x+5a} = \dfrac{x(y+7)+5(y+7)}{x(x+a)+5(x+a)} = \dfrac{(y+7)(x+5)}{(x+a)(x+5)} = \dfrac{y+7}{x+a}$

4. Performing the operations: $\dfrac{3x-12}{4} \cdot \dfrac{8}{2x-8} = \dfrac{3(x-4)}{4} \cdot \dfrac{8}{2(x-4)} = \dfrac{24(x-4)}{8(x-4)} = 3$

5. Performing the operations:

$$\frac{x^2-49}{x+1} \div \frac{x+7}{x^2-1} = \frac{x^2-49}{x+1} \cdot \frac{x^2-1}{x+7}$$
$$= \frac{(x+7)(x-7)}{x+1} \cdot \frac{(x+1)(x-1)}{x+7}$$
$$= \frac{(x+7)(x-7)(x+1)(x-1)}{(x+1)(x+7)}$$
$$= (x-7)(x-1)$$

6.   Performing the operations:

$$\frac{x^2-3x-10}{x^2-8x+15} \div \frac{3x^2+2x-8}{x^2+x-12} = \frac{x^2-3x-10}{x^2-8x+15} \cdot \frac{x^2+x-12}{3x^2+2x-8}$$

$$= \frac{(x-5)(x+2)}{(x-5)(x-3)} \cdot \frac{(x+4)(x-3)}{(3x-4)(x+2)}$$

$$= \frac{(x-5)(x+2)(x+4)(x-3)}{(x-5)(x-3)(3x-4)(x+2)}$$

$$= \frac{x+4}{3x-4}$$

7.   Performing the operations: $\left(x^2-9\right)\left(\dfrac{x+2}{x+3}\right) = \dfrac{(x+3)(x-3)}{1} \cdot \dfrac{x+2}{x+3} = \dfrac{(x+3)(x-3)(x+2)}{x+3} = (x-3)(x+2)$

8.   Performing the operations: $\dfrac{3}{x-2} - \dfrac{6}{x-2} = \dfrac{3-6}{x-2} = \dfrac{-3}{x-2}$

9.   Performing the operations:

$$\frac{x}{x^2-9} + \frac{4}{4x-12} = \frac{x}{(x+3)(x-3)} + \frac{4}{4(x-3)}$$

$$= \frac{x}{(x+3)(x-3)} + \frac{1 \cdot (x+3)}{(x+3)(x-3)}$$

$$= \frac{x}{(x+3)(x-3)} + \frac{x+3}{(x+3)(x-3)}$$

$$= \frac{2x+3}{(x+3)(x-3)}$$

10.   Performing the operations:

$$\frac{2x}{x^2-1} + \frac{x}{x^2-3x+2} = \frac{2x}{(x+1)(x-1)} + \frac{x}{(x-1)(x-2)}$$

$$= \frac{2x \cdot (x-2)}{(x+1)(x-1)(x-2)} + \frac{x \cdot (x+1)}{(x+1)(x-1)(x-2)}$$

$$= \frac{2x^2-4x}{(x+1)(x-1)(x-2)} + \frac{x^2+x}{(x+1)(x-1)(x-2)}$$

$$= \frac{3x^2-3x}{(x+1)(x-1)(x-2)}$$

$$= \frac{3x(x-1)}{(x+1)(x-1)(x-2)}$$

$$= \frac{3x}{(x+1)(x-2)}$$

11.   Multiplying both sides of the equation by 15:

$$15 \cdot \tfrac{7}{5} = 15 \cdot \frac{x+2}{3}$$

$$21 = 5(x+2)$$

$$21 = 5x+10$$

$$11 = 5x$$

$$x = \tfrac{11}{5}$$

Since $x = \tfrac{11}{5}$ checks in the original equation, the solution is $x = \tfrac{11}{5}$.

12.   Multiplying both sides of the equation by $x(x+4)$:

$$x(x+4) \cdot \frac{10}{x+4} = x(x+4) \cdot \left(\frac{6}{x} - \frac{4}{x} \right)$$
$$10x = 6(x+4) - 4(x+4)$$
$$10x = 6x + 24 - 4x - 16$$
$$10x = 2x + 8$$
$$8x = 8$$
$$x = 1$$

Since $x = 1$ checks in the original equation, the solution is $x = 1$.

13.   Multiplying both sides of the equation by $x^2 - x - 2 = (x-2)(x+1)$:

$$(x-2)(x+1)\left(\frac{3}{x-2} - \frac{4}{x+1} \right) = (x-2)(x+1) \cdot \frac{5}{(x-2)(x+1)}$$
$$3(x+1) - 4(x-2) = 5$$
$$3x + 3 - 4x + 8 = 5$$
$$-x + 11 = 5$$
$$-x = -6$$
$$x = 6$$

Since $x = 6$ checks in the original equation, the solution is $x = 6$.

14.   Let x represent the speed of the boat in still water. Completing the table:

	d	r	t
Upstream	26	$x-2$	$\dfrac{26}{x-2}$
Downstream	34	$x+2$	$\dfrac{34}{x+2}$

The equation is:

$$\frac{26}{x-2} = \frac{34}{x+2}$$
$$(x+2)(x-2) \cdot \frac{26}{x-2} = (x+2)(x-2) \cdot \frac{34}{x+2}$$
$$26(x+2) = 34(x-2)$$
$$26x + 52 = 34x - 68$$
$$-8x + 52 = -68$$
$$-8x = -120$$
$$x = 15$$

The speed of the boat in still water is 15 mph.

15.   Let t represent the time to empty the pool with both pipes open. The equation is:

$$\tfrac{1}{12} - \tfrac{1}{15} = \frac{1}{t}$$
$$60t\left(\tfrac{1}{12} - \tfrac{1}{15} \right) = 60t \cdot \frac{1}{t}$$
$$5t - 4t = 60$$
$$t = 60$$

It will take 60 hours to empty the pool with both pipes open.

16.   The ratio of alcohol to water is given by: $\dfrac{27 \text{ ml}}{54 \text{ ml}} = \tfrac{1}{2}$

The ratio of alcohol to total volume is given by: $\dfrac{27 \text{ ml}}{81 \text{ ml}} = \tfrac{1}{3}$

17.   Comparing defective parts to total parts, the proportion is:

$$\frac{8}{100} = \frac{x}{1650}$$
$$100x = 13200$$
$$x = 132$$

The machine can be expected to produce 132 defective parts.

18.   Simplifying the complex fraction: $\dfrac{1+\dfrac{1}{x}}{1-\dfrac{1}{x}} = \dfrac{\left(1+\dfrac{1}{x}\right)\bullet x}{\left(1-\dfrac{1}{x}\right)\bullet x} = \dfrac{x+1}{x-1}$

19.   Simplifying the complex fraction: $\dfrac{1-\dfrac{16}{x^2}}{1-\dfrac{2}{x}-\dfrac{8}{x^2}} = \dfrac{\left(1-\dfrac{16}{x^2}\right)\bullet x^2}{\left(1-\dfrac{2}{x}-\dfrac{8}{x^2}\right)\bullet x^2} = \dfrac{x^2-16}{x^2-2x-8} = \dfrac{(x+4)(x-4)}{(x-4)(x+2)} = \dfrac{x+4}{x+2}$

20.   The variation equation is $y = Kx^2$. Finding K:

$$36 = K\bullet 3^2$$
$$36 = 9K$$
$$K = 4$$

So $y = 4x^2$. Substituting $x = 5$: $y = 4\bullet 5^2 = 4\bullet 25 = 100$

21.   The variation equation is $y = \dfrac{K}{x}$. Finding K:

$$6 = \dfrac{K}{3}$$
$$K = 18$$

So $y = \dfrac{18}{x}$. Substituting $x = 9$: $y = \dfrac{18}{9} = 2$

Chapter 8
Roots and Radicals

8.1 Definitions and Common Roots

1. Finding the root: $\sqrt{9} = 3$

3. Finding the root: $-\sqrt{9} = -3$

5. Finding the root: $\sqrt{-25}$ is not a real number

7. Finding the root: $-\sqrt{144} = -12$

9. Finding the root: $\sqrt{625} = 25$

11. Finding the root: $\sqrt{-49}$ is not a real number

13. Finding the root: $-\sqrt{64} = -8$

15. Finding the root: $-\sqrt{100} = -10$

17. Finding the root: $\sqrt{1225} = 35$

19. Finding the root: $\sqrt[4]{1} = 1$

21. Finding the root: $\sqrt[3]{-8} = -2$

23. Finding the root: $-\sqrt[3]{125} = -5$

25. Finding the root: $\sqrt[3]{-1} = -1$

27. Finding the root: $\sqrt[3]{-27} = -3$

29. Finding the root: $-\sqrt[4]{16} = -2$

31. Simplifying the expression: $\sqrt{x^2} = x$

33. Simplifying the expression: $\sqrt{9x^2} = 3x$

35. Simplifying the expression: $\sqrt{x^2 y^2} = xy$

37. Simplifying the expression: $\sqrt{(a+b)^2} = a+b$

39. Simplifying the expression: $\sqrt{49x^2 y^2} = 7xy$

41. Simplifying the expression: $\sqrt[3]{x^3} = x$

43. Simplifying the expression: $\sqrt[3]{8x^3} = 2x$

45. Simplifying the expression: $\sqrt{x^4} = x^2$

47. Simplifying the expression: $\sqrt{36a^6} = 6a^3$

49. Simplifying the expression: $\sqrt{25a^8 b^4} = 5a^4 b^2$

51. Simplifying the expression: $\sqrt[3]{x^6} = x^2$

53. Simplifying the expression: $\sqrt[3]{27a^{12}} = 3a^4$

55. Simplifying the expression: $\sqrt[4]{x^8} = x^2$

57. Simplifying the expression: $\sqrt{9} + \sqrt{16} = 3 + 4 = 7$

59. Simplifying the expression: $\sqrt{9+16} = \sqrt{25} = 5$

61. Simplifying the expression: $\sqrt{144} + \sqrt{25} = 12 + 5 = 17$

63. Simplifying the expression: $\sqrt{144 + 25} = \sqrt{169} = 13$

65. Simplifying each expression:
$$\frac{5 + \sqrt{49}}{2} = \frac{5 + 7}{2} = \frac{12}{2} = 6 \qquad \frac{5 - \sqrt{49}}{2} = \frac{5 - 7}{2} = \frac{-2}{2} = -1$$

67. Simplifying each expression:
$$\frac{2 + \sqrt{16}}{2} = \frac{2 + 4}{2} = \frac{6}{2} = 3 \qquad \frac{2 - \sqrt{16}}{2} = \frac{2 - 4}{2} = \frac{-2}{2} = -1$$

69. Simplifying the expression: $\sqrt{x^2 + 6x + 9} = \sqrt{(x+3)^2} = x + 3$

71. Completing the table:

Length L (feet)	Time T (seconds)
1	1.11
2	1.57
3	1.92
4	2.22
5	2.48
6	2.72

73. Finding the annual rate of return: $r = \dfrac{\sqrt{65} - \sqrt{50}}{\sqrt{50}} \approx 0.140 = 14.0\%$

75. Finding the annual rate of return: $r = \dfrac{\sqrt{600} - \sqrt{500}}{\sqrt{500}} \approx 0.095 = 9.5\%$

77. Using the Pythagorean Theorem:
$$x^2 = 3^2 + 4^2 = 9 + 16 = 25$$
$$x = \sqrt{25} = 5$$

79. Using the Pythagorean Theorem:
$$x^2 = 5^2 + 10^2 = 25 + 100 = 125$$
$$x = \sqrt{125} \approx 11.2$$

81. Let l represent the length of the wire. Using the Pythagorean Theorem:
$$l^2 = 24^2 + 18^2 = 576 + 324 = 900$$
$$l = \sqrt{900} = 30$$
The length of the wire is 30 feet.

83. Let x represent the length of the log. Using the Pythagorean theorem:
$$x^2 = 5^2 + 12^2 = 25 + 144 = 169$$
$$x = \sqrt{169} = 13$$
The log must be 13 feet to just barely reach.

85. Reducing the rational expression: $\dfrac{x^2 - 16}{x + 4} = \dfrac{(x+4)(x-4)}{x+4} = x - 4$

87. Reducing the rational expression: $\dfrac{10a + 20}{5a^2 - 20} = \dfrac{10(a+2)}{5(a^2 - 4)} = \dfrac{10(a+2)}{5(a+2)(a-2)} = \dfrac{2}{a-2}$

89. Reducing the rational expression: $\dfrac{2x^2 - 5x - 3}{x^2 - 3x} = \dfrac{(2x+1)(x-3)}{x(x-3)} = \dfrac{2x+1}{x}$

91. Reducing the rational expression: $\dfrac{xy + 3x + 2y + 6}{xy + 3x + ay + 3a} = \dfrac{x(y+3) + 2(y+3)}{x(y+3) + a(y+3)} = \dfrac{(y+3)(x+2)}{(y+3)(x+a)} = \dfrac{x+2}{x+a}$

8.2 Properties of Radicals

1. Simplifying the radical expression: $\sqrt{8} = \sqrt{4 \cdot 2} = \sqrt{4}\sqrt{2} = 2\sqrt{2}$

3. Simplifying the radical expression: $\sqrt{12} = \sqrt{4 \cdot 3} = \sqrt{4}\sqrt{3} = 2\sqrt{3}$

5. Simplifying the radical expression: $\sqrt[3]{24} = \sqrt[3]{8 \cdot 3} = \sqrt[3]{8}\sqrt[3]{3} = 2\sqrt[3]{3}$

7. Simplifying the radical expression: $\sqrt{50x^2} = \sqrt{25x^2 \cdot 2} = \sqrt{25x^2}\sqrt{2} = 5x\sqrt{2}$

9. Simplifying the radical expression: $\sqrt{45a^2 b^2} = \sqrt{9a^2 b^2 \cdot 5} = \sqrt{9a^2 b^2}\sqrt{5} = 3ab\sqrt{5}$

11. Simplifying the radical expression: $\sqrt[3]{54x^3} = \sqrt[3]{27x^3 \cdot 2} = \sqrt[3]{27x^3}\sqrt[3]{2} = 3x\sqrt[3]{2}$

13. Simplifying the radical expression: $\sqrt{32x^4} = \sqrt{16x^4 \cdot 2} = \sqrt{16x^4}\sqrt{2} = 4x^2\sqrt{2}$

15. Simplifying the radical expression: $5\sqrt{80} = 5\sqrt{16 \cdot 5} = 5\sqrt{16}\sqrt{5} = 5 \cdot 4\sqrt{5} = 20\sqrt{5}$

17. Simplifying the radical expression: $\frac{1}{2}\sqrt{28x^3} = \frac{1}{2}\sqrt{4x^2 \cdot 7x} = \frac{1}{2}\sqrt{4x^2}\sqrt{7x} = \frac{1}{2} \cdot 2x\sqrt{7x} = x\sqrt{7x}$

19. Simplifying the radical expression: $x\sqrt[3]{8x^4} = x\sqrt[3]{8x^3 \cdot x} = x\sqrt[3]{8x^3}\sqrt[3]{x} = x \cdot 2x\sqrt[3]{x} = 2x^2\sqrt[3]{x}$

21. Simplifying the radical expression: $2a\sqrt[3]{27a^5} = 2a\sqrt[3]{27a^3 \cdot a^2} = 2a\sqrt[3]{27a^3}\sqrt[3]{a^2} = 2a \cdot 3a\sqrt[3]{a^2} = 6a^2\sqrt[3]{a^2}$

23. Simplifying the radical expression: $\frac{4}{3}\sqrt{45a^3} = \frac{4}{3}\sqrt{9a^2 \cdot 5a} = \frac{4}{3}\sqrt{9a^2}\sqrt{5a} = \frac{4}{3} \cdot 3a\sqrt{5a} = 4a\sqrt{5a}$

25. Simplifying the radical expression: $3\sqrt{50xy^2} = 3\sqrt{25y^2 \cdot 2x} = 3\sqrt{25y^2}\sqrt{2x} = 3 \cdot 5y\sqrt{2x} = 15y\sqrt{2x}$

27. Simplifying the radical expression: $7\sqrt{12x^2y} = 7\sqrt{4x^2 \cdot 3y} = 7\sqrt{4x^2}\sqrt{3y} = 7 \cdot 2x\sqrt{3y} = 14x\sqrt{3y}$

29. Simplifying the radical expression: $\sqrt{\frac{16}{25}} = \frac{\sqrt{16}}{\sqrt{25}} = \frac{4}{5}$ **31.** Simplifying the radical expression: $\sqrt{\frac{4}{9}} = \frac{\sqrt{4}}{\sqrt{9}} = \frac{2}{3}$

33. Simplifying the radical expression: $\sqrt[3]{\frac{8}{27}} = \frac{\sqrt[3]{8}}{\sqrt[3]{27}} = \frac{2}{3}$ **35.** Simplifying the radical expression: $\sqrt[4]{\frac{16}{81}} = \frac{\sqrt[4]{16}}{\sqrt[4]{81}} = \frac{2}{3}$

37. Simplifying the radical expression: $\sqrt{\frac{100x^2}{25}} = \sqrt{4x^2} = 2x$

39. Simplifying the radical expression: $\sqrt{\frac{81a^2b^2}{9}} = \sqrt{9a^2b^2} = 3ab$

41. Simplifying the radical expression: $\sqrt[3]{\frac{27x^3}{8y^3}} = \frac{\sqrt[3]{27x^3}}{\sqrt[3]{8y^3}} = \frac{3x}{2y}$

43. Simplifying the radical expression: $\sqrt{\frac{50}{9}} = \frac{\sqrt{50}}{\sqrt{9}} = \frac{\sqrt{25 \cdot 2}}{3} = \frac{5\sqrt{2}}{3}$

45. Simplifying the radical expression: $\sqrt{\frac{75}{25}} = \sqrt{3}$

47. Simplifying the radical expression: $\sqrt{\frac{128}{49}} = \frac{\sqrt{128}}{\sqrt{49}} = \frac{\sqrt{64 \cdot 2}}{7} = \frac{8\sqrt{2}}{7}$

49. Simplifying the radical expression: $\sqrt{\frac{288x}{25}} = \frac{\sqrt{288x}}{\sqrt{25}} = \frac{\sqrt{144 \cdot 2x}}{5} = \frac{12\sqrt{2x}}{5}$

51. Simplifying the radical expression: $\sqrt{\frac{54a^2}{25}} = \frac{\sqrt{54a^2}}{\sqrt{25}} = \frac{\sqrt{9a^2 \cdot 6}}{5} = \frac{3a\sqrt{6}}{5}$

53. Simplifying the radical expression: $\frac{3\sqrt{50}}{2} = \frac{3\sqrt{25 \cdot 2}}{2} = \frac{3 \cdot 5\sqrt{2}}{2} = \frac{15\sqrt{2}}{2}$

55. Simplifying the radical expression: $\frac{7\sqrt{28y^2}}{3} = \frac{7\sqrt{4y^2 \cdot 7}}{3} = \frac{7 \cdot 2y\sqrt{7}}{3} = \frac{14y\sqrt{7}}{3}$

57. Simplifying the radical expression: $\frac{5\sqrt{72a^2b^2}}{\sqrt{36}} = \frac{5\sqrt{36a^2b^2 \cdot 2}}{6} = \frac{5 \cdot 6ab\sqrt{2}}{6} = \frac{30ab\sqrt{2}}{6} = 5ab\sqrt{2}$

59. Simplifying the radical expression: $\frac{6\sqrt{8x^2y}}{\sqrt{4}} = \frac{6\sqrt{4x^2 \cdot 2y}}{2} = \frac{6 \cdot 2x\sqrt{2y}}{2} = \frac{12x\sqrt{2y}}{2} = 6x\sqrt{2y}$

61. Completing the table:

x	$\sqrt{x}$	$2\sqrt{x}$	$\sqrt{4x}$
1	1	2	2
2	1.414	2.828	2.828
3	1.732	3.464	3.464
4	2	4	4

63. Completing the table:

x	$\sqrt{x}$	$3\sqrt{x}$	$\sqrt{9x}$
1	1	3	3
2	1.414	4.243	4.243
3	1.732	5.196	5.196
4	2	6	6

65. Substituting $h = 25$ feet: $t = \sqrt{\frac{25}{16}} = \frac{\sqrt{25}}{\sqrt{16}} = \frac{5}{4} = 1\frac{1}{4}$ seconds

67. Performing the operations: $\frac{8x}{x^2 - 5x} \cdot \frac{x^2 - 25}{4x^2 + 4x} = \frac{8x}{x(x-5)} \cdot \frac{(x+5)(x-5)}{4x(x+1)} = \frac{8x(x+5)(x-5)}{4x^2(x-5)(x+1)} = \frac{2(x+5)}{x(x+1)}$

69. Performing the operations:
$$\frac{x^2+3x-4}{3x^2+7x-20} \div \frac{x^2-2x+1}{3x^2-2x-5} = \frac{x^2+3x-4}{3x^2+7x-20} \cdot \frac{3x^2-2x-5}{x^2-2x+1}$$
$$= \frac{(x+4)(x-1)}{(3x-5)(x+4)} \cdot \frac{(3x-5)(x+1)}{(x-1)^2}$$
$$= \frac{(x+4)(x-1)(3x-5)(x+1)}{(3x-5)(x+4)(x-1)^2}$$
$$= \frac{x+1}{x-1}$$

71. Performing the operations: $\left(x^2-36\right)\left(\frac{x+3}{x-6}\right) = \frac{(x+6)(x-6)}{1} \cdot \frac{x+3}{x-6} = \frac{(x+6)(x-6)(x+3)}{x-6} = (x+6)(x+3)$

8.3 Simplified Form for Radicals

1. Simplifying the radical expression: $\sqrt{\frac{1}{2}} = \frac{\sqrt{1}}{\sqrt{2}} \cdot \frac{\sqrt{2}}{\sqrt{2}} = \frac{\sqrt{2}}{\sqrt{4}} = \frac{\sqrt{2}}{2}$

3. Simplifying the radical expression: $\sqrt{\frac{1}{3}} = \frac{\sqrt{1}}{\sqrt{3}} \cdot \frac{\sqrt{3}}{\sqrt{3}} = \frac{\sqrt{3}}{\sqrt{9}} = \frac{\sqrt{3}}{3}$

5. Simplifying the radical expression: $\sqrt{\frac{2}{5}} = \frac{\sqrt{2}}{\sqrt{5}} \cdot \frac{\sqrt{5}}{\sqrt{5}} = \frac{\sqrt{10}}{\sqrt{25}} = \frac{\sqrt{10}}{5}$

7. Simplifying the radical expression: $\sqrt{\frac{3}{2}} = \frac{\sqrt{3}}{\sqrt{2}} \cdot \frac{\sqrt{2}}{\sqrt{2}} = \frac{\sqrt{6}}{\sqrt{4}} = \frac{\sqrt{6}}{2}$

9. Simplifying the radical expression: $\sqrt{\frac{20}{3}} = \frac{\sqrt{20}}{\sqrt{3}} \cdot \frac{\sqrt{3}}{\sqrt{3}} = \frac{\sqrt{60}}{\sqrt{9}} = \frac{\sqrt{4 \cdot 15}}{3} = \frac{2\sqrt{15}}{3}$

11. Simplifying the radical expression: $\sqrt{\frac{45}{6}} = \sqrt{\frac{15}{2}} = \frac{\sqrt{15}}{\sqrt{2}} \cdot \frac{\sqrt{2}}{\sqrt{2}} = \frac{\sqrt{30}}{\sqrt{4}} = \frac{\sqrt{30}}{2}$

13. Simplifying the radical expression: $\sqrt{\frac{20}{5}} = \sqrt{4} = 2$

15. Simplifying the radical expression: $\frac{\sqrt{21}}{\sqrt{3}} = \sqrt{\frac{21}{3}} = \sqrt{7}$

17. Simplifying the radical expression: $\frac{\sqrt{35}}{\sqrt{7}} = \sqrt{\frac{35}{7}} = \sqrt{5}$

19. Simplifying the radical expression: $\frac{10\sqrt{15}}{5\sqrt{3}} = \frac{10}{5} \cdot \sqrt{\frac{15}{3}} = 2\sqrt{5}$

21. Simplifying the radical expression: $\frac{6\sqrt{21}}{3\sqrt{7}} = \frac{6}{3} \cdot \sqrt{\frac{21}{7}} = 2\sqrt{3}$

23. Simplifying the radical expression: $\frac{6\sqrt{35}}{12\sqrt{5}} = \frac{6}{12} \cdot \sqrt{\frac{35}{5}} = \frac{1}{2}\sqrt{7} = \frac{\sqrt{7}}{2}$

25. Simplifying the radical expression: $\sqrt{\frac{4x^2y^2}{2}} = \sqrt{2x^2y^2} = xy\sqrt{2}$

27. Simplifying the radical expression: $\sqrt{\frac{5x^2y}{3}} = \frac{\sqrt{5x^2y}}{\sqrt{3}} \cdot \frac{\sqrt{3}}{\sqrt{3}} = \frac{\sqrt{15x^2y}}{\sqrt{9}} = \frac{x\sqrt{15y}}{3}$

29. Simplifying the radical expression: $\sqrt{\frac{16a^4}{5}} = \frac{\sqrt{16a^4}}{\sqrt{5}} \cdot \frac{\sqrt{5}}{\sqrt{5}} = \frac{4a^2\sqrt{5}}{\sqrt{25}} = \frac{4a^2\sqrt{5}}{5}$

31. Simplifying the radical expression: $\sqrt{\frac{72a^5}{5}} = \frac{\sqrt{72a^5}}{\sqrt{5}} \cdot \frac{\sqrt{5}}{\sqrt{5}} = \frac{\sqrt{360a^5}}{\sqrt{25}} = \frac{\sqrt{36a^4 \cdot 10a}}{5} = \frac{6a^2\sqrt{10a}}{5}$

33. Simplifying the radical expression: $\sqrt{\dfrac{20x^2y^3}{3}} = \dfrac{\sqrt{20x^2y^3}}{\sqrt{3}} \cdot \dfrac{\sqrt{3}}{\sqrt{3}} = \dfrac{\sqrt{60x^2y^3}}{\sqrt{9}} = \dfrac{\sqrt{4x^2y^2 \cdot 15y}}{3} = \dfrac{2xy\sqrt{15y}}{3}$

35. Simplifying the radical expression: $\dfrac{2\sqrt{20x^2y^3}}{3} = \dfrac{2\sqrt{4x^2y^2 \cdot 5y}}{3} = \dfrac{2 \cdot 2xy\sqrt{5y}}{3} = \dfrac{4xy\sqrt{5y}}{3}$

37. Simplifying the radical expression: $\dfrac{6\sqrt{54a^2b^3}}{5} = \dfrac{6\sqrt{9a^2b^2 \cdot 6b}}{5} = \dfrac{6 \cdot 3ab\sqrt{6b}}{5} = \dfrac{18ab\sqrt{6b}}{5}$

39. Simplifying the radical expression: $\dfrac{3\sqrt{72x^4}}{\sqrt{2x}} = 3\sqrt{\dfrac{72x^4}{2x}} = 3\sqrt{36x^3} = 3\sqrt{36x^2 \cdot x} = 3 \cdot 6x\sqrt{x} = 18x\sqrt{x}$

41. Simplifying the radical expression: $\sqrt[3]{\dfrac{1}{2}} = \dfrac{\sqrt[3]{1}}{\sqrt[3]{2}} \cdot \dfrac{\sqrt[3]{4}}{\sqrt[3]{4}} = \dfrac{\sqrt[3]{4}}{\sqrt[3]{8}} = \dfrac{\sqrt[3]{4}}{2}$

43. Simplifying the radical expression: $\sqrt[3]{\dfrac{1}{9}} = \dfrac{\sqrt[3]{1}}{\sqrt[3]{9}} \cdot \dfrac{\sqrt[3]{3}}{\sqrt[3]{3}} = \dfrac{\sqrt[3]{3}}{\sqrt[3]{27}} = \dfrac{\sqrt[3]{3}}{3}$

45. Simplifying the radical expression: $\sqrt[3]{\dfrac{3}{2}} = \dfrac{\sqrt[3]{3}}{\sqrt[3]{2}} \cdot \dfrac{\sqrt[3]{4}}{\sqrt[3]{4}} = \dfrac{\sqrt[3]{12}}{\sqrt[3]{8}} = \dfrac{\sqrt[3]{12}}{2}$

47. Completing the table:

x	$\sqrt{x}$	$\dfrac{1}{\sqrt{x}}$	$\dfrac{\sqrt{x}}{x}$
1	1	1	1
2	1.414	0.707	0.707
3	1.732	0.577	0.577
4	2	0.5	0.5
5	2.236	0.447	0.447
6	2.449	0.408	0.408

49. Completing the table:

x	$\sqrt{x^2}$	$\sqrt{x^3}$	$x\sqrt{x}$
1	1	1	1
2	2	2.828	2.828
3	3	5.196	5.196
4	4	8	8
5	5	11.180	11.180
6	6	14.697	14.697

51. Substituting $h = 24$: $d = \sqrt{\dfrac{3 \cdot 24}{2}} = \sqrt{36} = 6$ miles

53. Drawing the figure:

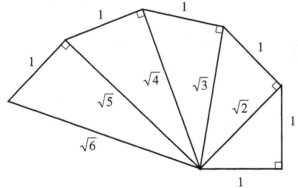

55. Simplifying the terms in the sequence:
$$\sqrt{1^2 + 1} = \sqrt{1 + 1} = \sqrt{2}$$
$$\sqrt{\left(\sqrt{2}\right)^2 + 1} = \sqrt{2 + 1} = \sqrt{3}$$
$$\sqrt{\left(\sqrt{3}\right)^2 + 1} = \sqrt{3 + 1} = \sqrt{4} = 2$$

57. Combining the terms: $3x + 7x = (3 + 7)x = 10x$

59. Combining the terms: $15x + 8x = (15 + 8)x = 23x$

61. Combining the terms: $7a - 3a + 6a = (7 - 3 + 6)a = 10a$

63. Combining the rational expressions: $\dfrac{x^2}{x+5} + \dfrac{10x+25}{x+5} = \dfrac{x^2+10x+25}{x+5} = \dfrac{(x+5)^2}{x+5} = x+5$

65. Combining the rational expressions: $\dfrac{a}{3} + \dfrac{2}{5} = \dfrac{a \bullet 5}{3 \bullet 5} + \dfrac{2 \bullet 3}{5 \bullet 3} = \dfrac{5a}{15} + \dfrac{6}{15} = \dfrac{5a+6}{15}$

67. Combining the rational expressions:

$$
\begin{aligned}
\dfrac{6}{a^2-9} - \dfrac{5}{a^2-a-6} &= \dfrac{6}{(a+3)(a-3)} - \dfrac{5}{(a-3)(a+2)} \\[2mm]
&= \dfrac{6 \bullet (a+2)}{(a+3)(a-3)(a+2)} - \dfrac{5 \bullet (a+3)}{(a+3)(a-3)(a+2)} \\[2mm]
&= \dfrac{6a+12}{(a+3)(a-3)(a+2)} - \dfrac{5a+15}{(a+3)(a-3)(a+2)} \\[2mm]
&= \dfrac{6a+12-5a-15}{(a+3)(a-3)(a+2)} \\[2mm]
&= \dfrac{a-3}{(a+3)(a-3)(a+2)} \\[2mm]
&= \dfrac{1}{(a+3)(a+2)}
\end{aligned}
$$

8.4 Addition and Subtraction of Radical Expressions

1. Simplifying the radical expression: $3\sqrt{2} + 4\sqrt{2} = 7\sqrt{2}$

3. Simplifying the radical expression: $9\sqrt{5} - 7\sqrt{5} = 2\sqrt{5}$

5. Simplifying the radical expression: $\sqrt{3} + 6\sqrt{3} = 7\sqrt{3}$

7. Simplifying the radical expression: $\frac{5}{8}\sqrt{5} - \frac{3}{7}\sqrt{5} = \frac{35}{56}\sqrt{5} - \frac{24}{56}\sqrt{5} = \frac{11}{56}\sqrt{5}$

9. Simplifying the radical expression: $14\sqrt{13} - \sqrt{13} = 13\sqrt{13}$

11. Simplifying the radical expression: $-3\sqrt{10} + 9\sqrt{10} = 6\sqrt{10}$

13. Simplifying the radical expression: $5\sqrt{5} + \sqrt{5} = 6\sqrt{5}$

15. Simplifying the radical expression: $\sqrt{8} + 2\sqrt{2} = \sqrt{4 \bullet 2} + 2\sqrt{2} = 2\sqrt{2} + 2\sqrt{2} = 4\sqrt{2}$

17. Simplifying the radical expression: $3\sqrt{3} - \sqrt{27} = 3\sqrt{3} - \sqrt{9 \bullet 3} = 3\sqrt{3} - 3\sqrt{3} = 0$

19. Simplifying the radical expression:
$$5\sqrt{12} - 10\sqrt{48} = 5\sqrt{4 \bullet 3} - 10\sqrt{16 \bullet 3} = 5 \bullet 2\sqrt{3} - 10 \bullet 4\sqrt{3} = 10\sqrt{3} - 40\sqrt{3} = -30\sqrt{3}$$

21. Simplifying the radical expression: $-\sqrt{75} - \sqrt{3} = -\sqrt{25 \bullet 3} - \sqrt{3} = -5\sqrt{3} - \sqrt{3} = -6\sqrt{3}$

23. Simplifying the radical expression: $\frac{1}{5}\sqrt{75} - \frac{1}{2}\sqrt{12} = \frac{1}{5}\sqrt{25 \bullet 3} - \frac{1}{2}\sqrt{4 \bullet 3} = \frac{1}{5} \bullet 5\sqrt{3} - \frac{1}{2} \bullet 2\sqrt{3} = \sqrt{3} - \sqrt{3} = 0$

25. Simplifying the radical expression: $\frac{3}{4}\sqrt{8} + \frac{3}{10}\sqrt{75} = \frac{3}{4}\sqrt{4 \bullet 2} + \frac{3}{10}\sqrt{25 \bullet 3} = \frac{3}{4} \bullet 2\sqrt{2} + \frac{3}{10} \bullet 5\sqrt{3} = \frac{3}{2}\sqrt{2} + \frac{3}{2}\sqrt{3}$

27. Simplifying the radical expression: $\sqrt{27} - 2\sqrt{12} + \sqrt{3} = \sqrt{9 \bullet 3} - 2\sqrt{4 \bullet 3} + \sqrt{3} = 3\sqrt{3} - 4\sqrt{3} + \sqrt{3} = 0$

29. Simplifying the radical expression:
$$
\begin{aligned}
\tfrac{5}{6}\sqrt{72} - \tfrac{3}{8}\sqrt{8} + \tfrac{3}{10}\sqrt{50} &= \tfrac{5}{6}\sqrt{36 \bullet 2} - \tfrac{3}{8}\sqrt{4 \bullet 2} + \tfrac{3}{10}\sqrt{25 \bullet 2} \\[1mm]
&= \tfrac{5}{6} \bullet 6\sqrt{2} - \tfrac{3}{8} \bullet 2\sqrt{2} + \tfrac{3}{10} \bullet 5\sqrt{2} \\[1mm]
&= 5\sqrt{2} - \tfrac{3}{4}\sqrt{2} + \tfrac{3}{2}\sqrt{2} \\[1mm]
&= \tfrac{20}{4}\sqrt{2} - \tfrac{3}{4}\sqrt{2} + \tfrac{6}{4}\sqrt{2} \\[1mm]
&= \tfrac{23}{4}\sqrt{2}
\end{aligned}
$$

31. Simplifying the radical expression:
$$5\sqrt{7} + 2\sqrt{28} - 4\sqrt{63} = 5\sqrt{7} + 2\sqrt{4 \bullet 7} - 4\sqrt{9 \bullet 7} = 5\sqrt{7} + 2 \bullet 2\sqrt{7} - 4 \bullet 3\sqrt{7} = 5\sqrt{7} + 4\sqrt{7} - 12\sqrt{7} = -3\sqrt{7}$$

33.   Simplifying the radical expression:
$$6\sqrt{48} - 2\sqrt{12} + 5\sqrt{27} = 6\sqrt{16 \cdot 3} - 2\sqrt{4 \cdot 3} + 5\sqrt{9 \cdot 3}$$
$$= 6 \cdot 4\sqrt{3} - 2 \cdot 2\sqrt{3} + 5 \cdot 3\sqrt{3}$$
$$= 24\sqrt{3} - 4\sqrt{3} + 15\sqrt{3}$$
$$= 35\sqrt{3}$$

35.   Simplifying the radical expression:
$$6\sqrt{48} - \sqrt{72} - 3\sqrt{300} = 6\sqrt{16 \cdot 3} - \sqrt{36 \cdot 2} - 3\sqrt{100 \cdot 3}$$
$$= 6 \cdot 4\sqrt{3} - 6\sqrt{2} - 3 \cdot 10\sqrt{3}$$
$$= 24\sqrt{3} - 6\sqrt{2} - 30\sqrt{3}$$
$$= -6\sqrt{3} - 6\sqrt{2}$$

37.   Simplifying the radical expression: $\sqrt{x^3} + x\sqrt{x} = \sqrt{x^2 \cdot x} + x\sqrt{x} = x\sqrt{x} + x\sqrt{x} = 2x\sqrt{x}$

39.   Simplifying the radical expression: $5\sqrt{3a^2} - a\sqrt{3} = 5a\sqrt{3} - a\sqrt{3} = 4a\sqrt{3}$

41.   Simplifying the radical expression:
$$5\sqrt{8x^3} + x\sqrt{50x} = 5\sqrt{4x^2 \cdot 2x} + x\sqrt{25 \cdot 2x} = 5 \cdot 2x\sqrt{2x} + x \cdot 5\sqrt{2x} = 10x\sqrt{2x} + 5x\sqrt{2x} = 15x\sqrt{2x}$$

43.   Simplifying the radical expression:
$$3\sqrt{75x^3 y} - 2x\sqrt{3xy} = 3\sqrt{25x^2 \cdot 3xy} - 2x\sqrt{3xy} = 3 \cdot 5x\sqrt{3xy} - 2x\sqrt{3xy} = 15x\sqrt{3xy} - 2x\sqrt{3xy} = 13x\sqrt{3xy}$$

45.   Simplifying the radical expression: $\sqrt{20ab^2} - b\sqrt{45a} = \sqrt{4b^2 \cdot 5a} - b\sqrt{9 \cdot 5a} = 2b\sqrt{5a} - 3b\sqrt{5a} = -b\sqrt{5a}$

47.   Simplifying the radical expression:
$$9\sqrt{18x^3} - 2x\sqrt{48x} = 9\sqrt{9x^2 \cdot 2x} - 2x\sqrt{16 \cdot 3x} = 9 \cdot 3x\sqrt{2x} - 2x \cdot 4\sqrt{3x} = 27x\sqrt{2x} - 8x\sqrt{3x}$$

49.   Simplifying the radical expression:
$$7\sqrt{50x^2 y} + 8x\sqrt{8y} - 7\sqrt{32x^2 y} = 7\sqrt{25x^2 \cdot 2y} + 8x\sqrt{4 \cdot 2y} - 7\sqrt{16x^2 \cdot 2y}$$
$$= 7 \cdot 5x\sqrt{2y} + 8x \cdot 2\sqrt{2y} - 7 \cdot 4x\sqrt{2y}$$
$$= 35x\sqrt{2y} + 16x\sqrt{2y} - 28x\sqrt{2y}$$
$$= 23x\sqrt{2y}$$

51.   Simplifying the expression: $\dfrac{8 - \sqrt{24}}{6} = \dfrac{8 - \sqrt{4 \cdot 6}}{6} = \dfrac{8 - 2\sqrt{6}}{6} = \dfrac{2\left(4 - \sqrt{6}\right)}{6} = \dfrac{4 - \sqrt{6}}{3}$

53.   Simplifying the expression: $\dfrac{6 + \sqrt{8}}{2} = \dfrac{6 + \sqrt{4 \cdot 2}}{2} = \dfrac{6 + 2\sqrt{2}}{2} = \dfrac{2\left(3 + \sqrt{2}\right)}{2} = 3 + \sqrt{2}$

55.   Simplifying the expression: $\dfrac{-10 + \sqrt{50}}{10} = \dfrac{-10 + \sqrt{25 \cdot 2}}{10} = \dfrac{-10 + 5\sqrt{2}}{10} = \dfrac{5\left(-2 + \sqrt{2}\right)}{10} = \dfrac{-2 + \sqrt{2}}{2}$

57.   Completing the table:

x	$\sqrt{x^2 + 9}$	$x + 3$
1	3.162	4
2	3.606	5
3	4.243	6
4	5	7
5	5.831	8
6	6.708	9

59.   Completing the table:

x	$\sqrt{x + 3}$	$\sqrt{x} + \sqrt{3}$
1	2	2.732
2	2.236	3.146
3	2.449	3.464
4	2.646	3.732
5	2.828	3.968
6	3	4.182

61.   The correct statement is: $4\sqrt{3} + 5\sqrt{3} = 9\sqrt{3}$

63.   Multiplying the expressions: $(3x + y)^2 = (3x)^2 + 2(3x)(y) + y^2 = 9x^2 + 6xy + y^2$

65.   Multiplying the expressions: $(3x - 4y)(3x + 4y) = (3x)^2 - (4y)^2 = 9x^2 - 16y^2$

67.   Multiplying each side of the equation by 6:
$$6\left(\frac{x}{3}-\frac{1}{2}\right)=6\left(\frac{5}{2}\right)$$
$$2x-3=15$$
$$2x=18$$
$$x=9$$
Since $x=9$ checks in the original equation, the solution is $x=9$.

69.   Multiplying each side of the equation by x^2:
$$x^2\left(1-\frac{5}{x}\right)=x^2\left(\frac{-6}{x^2}\right)$$
$$x^2-5x=-6$$
$$x^2-5x+6=0$$
$$(x-2)(x-3)=0$$
$$x=2,3$$
Both $x=2$ and $x=3$ check in the original equation.

71.   Multiplying each side of the equation by $2(a-4)$:
$$2(a-4)\left(\frac{a}{a-4}-\frac{a}{2}\right)=2(a-4)\cdot\frac{4}{a-4}$$
$$2a-a(a-4)=8$$
$$2a-a^2+4a=8$$
$$-a^2+6a-8=0$$
$$a^2-6a+8=0$$
$$(a-4)(a-2)=0$$
$$a=2,4$$
Since $a=4$ does not check in the original equation, the solution is $a=2$.

8.5 Multiplication and Division of Radicals

1.   Multiplying the radicals: $\sqrt{3}\sqrt{2}=\sqrt{6}$

3.   Multiplying the radicals: $\sqrt{6}\sqrt{2}=\sqrt{12}=\sqrt{4\cdot3}=2\sqrt{3}$

5.   Multiplying the radicals: $\left(2\sqrt{3}\right)\left(5\sqrt{7}\right)=10\sqrt{21}$

7.   Multiplying the radicals: $\left(4\sqrt{3}\right)\left(2\sqrt{6}\right)=8\sqrt{18}=8\sqrt{9\cdot2}=8\cdot3\sqrt{2}=24\sqrt{2}$

9.   Multiplying the radicals: $\sqrt{2}\left(\sqrt{3}-1\right)=\sqrt{6}-\sqrt{2}$

11.   Multiplying the radicals: $\sqrt{2}\left(\sqrt{3}+\sqrt{2}\right)=\sqrt{6}+\sqrt{4}=\sqrt{6}+2$

13.   Multiplying the radicals: $\sqrt{3}\left(2\sqrt{2}+\sqrt{3}\right)=2\sqrt{6}+\sqrt{9}=2\sqrt{6}+3$

15.   Multiplying the radicals: $\sqrt{3}\left(2\sqrt{3}-\sqrt{5}\right)=2\sqrt{9}-\sqrt{15}=2\cdot3-\sqrt{15}=6-\sqrt{15}$

17.   Multiplying the radicals: $2\sqrt{3}\left(\sqrt{2}+\sqrt{5}\right)=2\sqrt{6}+2\sqrt{15}$

19.   Simplifying the expression: $\left(\sqrt{2}+1\right)^2=\left(\sqrt{2}\right)^2+2\left(\sqrt{2}\right)(1)+(1)^2=2+2\sqrt{2}+1=3+2\sqrt{2}$

21.   Simplifying the expression: $\left(\sqrt{x}+3\right)^2=\left(\sqrt{x}\right)^2+2\left(\sqrt{x}\right)(3)+(3)^2=x+6\sqrt{x}+9$

23.   Simplifying the expression: $\left(5-\sqrt{2}\right)^2=(5)^2-2(5)\left(\sqrt{2}\right)+\left(\sqrt{2}\right)^2=25-10\sqrt{2}+2=27-10\sqrt{2}$

25.   Simplifying the expression: $\left(\sqrt{a}-\frac{1}{2}\right)^2=\left(\sqrt{a}\right)^2-2\left(\sqrt{a}\right)\left(\frac{1}{2}\right)+\left(\frac{1}{2}\right)^2=a-\sqrt{a}+\frac{1}{4}$

27.   Simplifying the expression: $\left(3+\sqrt{7}\right)^2=(3)^2+2(3)\left(\sqrt{7}\right)+\left(\sqrt{7}\right)^2=9+6\sqrt{7}+7=16+6\sqrt{7}$

29.   Simplifying the expression: $\left(\sqrt{5}+3\right)\left(\sqrt{5}+2\right)=5+3\sqrt{5}+2\sqrt{5}+6=11+5\sqrt{5}$

31. Simplifying the expression: $\left(\sqrt{2}-5\right)\left(\sqrt{2}+6\right)=2-5\sqrt{2}+6\sqrt{2}-30=-28+\sqrt{2}$

33. Simplifying the expression: $\left(\sqrt{3}+\frac{1}{2}\right)\left(\sqrt{2}+\frac{1}{3}\right)=\sqrt{6}+\frac{1}{3}\sqrt{2}+\frac{1}{2}\sqrt{3}+\frac{1}{6}$

35. Simplifying the expression: $\left(\sqrt{x}+6\right)\left(\sqrt{x}-6\right)=\left(\sqrt{x}\right)^2-\left(6\right)^2=x-36$

37. Simplifying the expression: $\left(\sqrt{a}+\frac{1}{3}\right)\left(\sqrt{a}+\frac{2}{3}\right)=a+\frac{1}{3}\sqrt{a}+\frac{2}{3}\sqrt{a}+\frac{2}{9}=a+\sqrt{a}+\frac{2}{9}$

39. Simplifying the expression: $\left(\sqrt{5}-2\right)\left(\sqrt{5}+2\right)=\left(\sqrt{5}\right)^2-\left(2\right)^2=5-4=1$

41. Simplifying the expression: $\left(2\sqrt{7}+3\right)\left(3\sqrt{7}-4\right)=6\sqrt{49}+9\sqrt{7}-8\sqrt{7}-12=42+\sqrt{7}-12=30+\sqrt{7}$

43. Simplifying the expression: $\left(2\sqrt{x}+4\right)\left(3\sqrt{x}+2\right)=6x+12\sqrt{x}+4\sqrt{x}+8=6x+16\sqrt{x}+8$

45. Simplifying the expression: $\left(7\sqrt{a}+2\sqrt{b}\right)\left(7\sqrt{a}-2\sqrt{b}\right)=\left(7\sqrt{a}\right)^2-\left(2\sqrt{b}\right)^2=49a-4b$

47. Rationalizing the denominator: $\dfrac{\sqrt{3}}{\sqrt{5}-\sqrt{2}}\cdot\dfrac{\sqrt{5}+\sqrt{2}}{\sqrt{5}+\sqrt{2}}=\dfrac{\sqrt{15}+\sqrt{6}}{5-2}=\dfrac{\sqrt{15}+\sqrt{6}}{3}$

49. Rationalizing the denominator: $\dfrac{\sqrt{5}}{\sqrt{5}+\sqrt{2}}\cdot\dfrac{\sqrt{5}-\sqrt{2}}{\sqrt{5}-\sqrt{2}}=\dfrac{\sqrt{25}-\sqrt{10}}{5-2}=\dfrac{5-\sqrt{10}}{3}$

51. Rationalizing the denominator: $\dfrac{8}{3-\sqrt{5}}\cdot\dfrac{3+\sqrt{5}}{3+\sqrt{5}}=\dfrac{8\left(3+\sqrt{5}\right)}{9-5}=\dfrac{8\left(3+\sqrt{5}\right)}{4}=2\left(3+\sqrt{5}\right)=6+2\sqrt{5}$

53. Rationalizing the denominator: $\dfrac{\sqrt{3}+\sqrt{2}}{\sqrt{3}-\sqrt{2}}\cdot\dfrac{\sqrt{3}+\sqrt{2}}{\sqrt{3}+\sqrt{2}}=\dfrac{3+\sqrt{6}+\sqrt{6}+2}{3-2}=\dfrac{5+2\sqrt{6}}{1}=5+2\sqrt{6}$

55. Rationalizing the denominator: $\dfrac{\sqrt{7}-\sqrt{3}}{\sqrt{7}+\sqrt{3}}\cdot\dfrac{\sqrt{7}-\sqrt{3}}{\sqrt{7}-\sqrt{3}}=\dfrac{7-\sqrt{21}-\sqrt{21}+3}{7-3}=\dfrac{10-2\sqrt{21}}{4}=\dfrac{2\left(5-\sqrt{21}\right)}{4}=\dfrac{5-\sqrt{21}}{2}$

57. Rationalizing the denominator: $\dfrac{\sqrt{x}+2}{\sqrt{x}-2}\cdot\dfrac{\sqrt{x}+2}{\sqrt{x}+2}=\dfrac{x+2\sqrt{x}+2\sqrt{x}+4}{x-4}=\dfrac{x+4\sqrt{x}+4}{x-4}$

59. Rationalizing the denominator: $\dfrac{\sqrt{5}-\sqrt{2}}{\sqrt{5}+\sqrt{3}}\cdot\dfrac{\sqrt{5}-\sqrt{3}}{\sqrt{5}-\sqrt{3}}=\dfrac{5-\sqrt{10}-\sqrt{15}+\sqrt{6}}{5-3}=\dfrac{5-\sqrt{10}-\sqrt{15}+\sqrt{6}}{2}$

61. The correct statement is: $2\left(3\sqrt{5}\right)=6\sqrt{5}$

63. The correct statement is: $\left(\sqrt{3}+7\right)^2=\left(\sqrt{3}\right)^2+2\left(\sqrt{3}\right)(7)+(7)^2=3+14\sqrt{3}+49=52+14\sqrt{3}$

65. Solving the equation by factoring:
$$x^2+5x-6=0$$
$$(x+6)(x-1)=0$$
$$x=-6,1$$

67. Solving the equation by factoring:
$$x^2-3x=0$$
$$x(x-3)=0$$
$$x=0,3$$

69. Solving the proportion:
$$\frac{x}{3}=\frac{27}{x}$$
$$x^2=81$$
$$x^2-81=0$$
$$(x+9)(x-9)=0$$
$$x=-9,9$$

71. Solving the proportion:
$$\frac{x}{5}=\frac{3}{x+2}$$
$$x^2+2x=15$$
$$x^2+2x-15=0$$
$$(x+5)(x-3)=0$$
$$x=-5,3$$

73. Comparing miles to hours, the proportion is:
$$\frac{375}{15}=\frac{x}{20}$$
$$15x=7500$$
$$x=500$$
You will drive 500 miles.

8.6 Equations Involving Radicals

1. Solving the equation by squaring:
$$\sqrt{x+1} = 2$$
$$\left(\sqrt{x+1}\right)^2 = (2)^2$$
$$x+1 = 4$$
$$x = 3$$
This value checks in the original equation.

3. Solving the equation by squaring:
$$\sqrt{x+5} = 7$$
$$\left(\sqrt{x+5}\right)^2 = (7)^2$$
$$x+5 = 49$$
$$x = 44$$
This value checks in the original equation.

5. Solving the equation by squaring:
$$\sqrt{x-9} = -6$$
$$\left(\sqrt{x-9}\right)^2 = (-6)^2$$
$$x-9 = 36$$
$$x = 45$$
Since $\sqrt{45-9} = \sqrt{36} = 6 \neq -6$, there is no solution to the equation.

7. Solving the equation by squaring:
$$\sqrt{x-5} = -4$$
$$\left(\sqrt{x-5}\right)^2 = (-4)^2$$
$$x-5 = 16$$
$$x = 21$$
Since $\sqrt{21-5} = \sqrt{16} = 4 \neq -4$, there is no solution to the equation.

9. Solving the equation by squaring:
$$\sqrt{x-8} = 0$$
$$\left(\sqrt{x-8}\right)^2 = (0)^2$$
$$x-8 = 0$$
$$x = 8$$
This value checks in the original equation.

11. Solving the equation by squaring:
$$\sqrt{2x+1} = 3$$
$$\left(\sqrt{2x+1}\right)^2 = (3)^2$$
$$2x+1 = 9$$
$$2x = 8$$
$$x = 4$$
This value checks in the original equation.

13. Solving the equation by squaring:
$$\sqrt{2x-3} = -5$$
$$\left(\sqrt{2x-3}\right)^2 = (-5)^2$$
$$2x-3 = 25$$
$$2x = 28$$
$$x = 14$$
Since $\sqrt{2\cdot14-3} = \sqrt{28-3} = \sqrt{25} = 5 \neq -5$, there is no solution to the equation.

15. Solving the equation by squaring:
$$\sqrt{3x+6} = 2$$
$$\left(\sqrt{3x+6}\right)^2 = (2)^2$$
$$3x+6 = 4$$
$$3x = -2$$
$$x = -\tfrac{2}{3}$$
This value checks in the original equation.

17. Solving the equation by squaring:
$$2\sqrt{x} = 10$$
$$\sqrt{x} = 5$$
$$\left(\sqrt{x}\right)^2 = (5)^2$$
$$x = 25$$
This value checks in the original equation.

19. Solving the equation by squaring:

$$3\sqrt{a} = 6$$
$$\sqrt{a} = 2$$
$$\left(\sqrt{a}\right)^2 = (2)^2$$
$$a = 4$$

This value checks in the original equation.

21. Solving the equation by squaring:
$$\sqrt{3x+4} - 3 = 2$$
$$\sqrt{3x+4} = 5$$
$$\left(\sqrt{3x+4}\right)^2 = (5)^2$$
$$3x+4 = 25$$
$$3x = 21$$
$$x = 7$$
This value checks in the original equation.

23. Solving the equation by squaring:
$$\sqrt{5y-4} - 2 = 4$$
$$\sqrt{5y-4} = 6$$
$$\left(\sqrt{5y-4}\right)^2 = (6)^2$$
$$5y-4 = 36$$
$$5y = 40$$
$$y = 8$$

This value checks in the original equation.

25. Solving the equation by squaring:
$$\sqrt{2x+1} + 5 = 2$$
$$\sqrt{2x+1} = -3$$
$$\left(\sqrt{2x+1}\right)^2 = (-3)^2$$
$$2x+1 = 9$$
$$2x = 8$$
$$x = 4$$

Since $\sqrt{2 \cdot 4 + 1} + 5 = \sqrt{9} + 5 = 3 + 5 = 8 \neq 2$, there is no solution to the equation.

27. Solving the equation by squaring:
$$\sqrt{x+3} = x-3$$
$$\left(\sqrt{x+3}\right)^2 = (x-3)^2$$
$$x+3 = x^2 - 6x + 9$$
$$0 = x^2 - 7x + 6$$
$$0 = (x-6)(x-1)$$
$$x = 6, 1$$

Since $x = 1$ does not check in the original equation, the solution is $x = 6$.

29. Solving the equation by squaring:
$$\sqrt{a+2} = a+2$$
$$\left(\sqrt{a+2}\right)^2 = (a+2)^2$$
$$a+2 = a^2 + 4a + 4$$
$$0 = a^2 + 3a + 2$$
$$0 = (a+2)(a+1)$$
$$a = -2, -1$$

Both values check in the original equation.

31. Solving the equation by squaring:
$$\sqrt{2x+9} = x+5$$
$$\left(\sqrt{2x+9}\right)^2 = (x+5)^2$$
$$2x+9 = x^2 + 10x + 25$$
$$0 = x^2 + 8x + 16$$
$$0 = (x+4)^2$$
$$x = -4$$

This value checks in the original equation.

33.   Solving the equation by squaring:

$$\sqrt{y-4} = y-6$$
$$\left(\sqrt{y-4}\right)^2 = (y-6)^2$$
$$y-4 = y^2 - 12y + 36$$
$$0 = y^2 - 13y + 40$$
$$0 = (y-8)(y-5)$$
$$y = 5, 8$$

Since $y = 5$ does not check in the original equation, the solution is $y = 8$.

35.   Drawing a line graph:

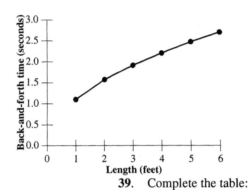

37.   Complete the table:

x	y
0	0
1	1
2	1.4
3	1.7
4	2

Sketching the graph:

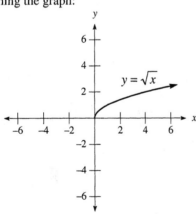

39.   Complete the table:

x	y
0	0
1	2
4	4
9	6

Sketching the graph:

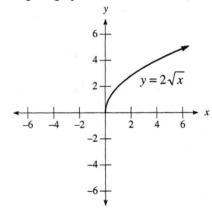

41.   Complete the table:

x	y
0	2
1	3
2	3.4
4	4
9	5

Sketching the graph:

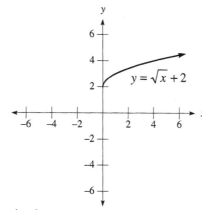

43.   Let x represent the number. The equation is:

$$x + 2 = \sqrt{8x}$$
$$(x + 2)^2 = \left(\sqrt{8x}\right)^2$$
$$x^2 + 4x + 4 = 8x$$
$$x^2 - 4x + 4 = 0$$
$$(x - 2)^2 = 0$$
$$x - 2 = 0$$
$$x = 2$$

The number is 2.

45.   Let x represent the number. The equation is:

$$x - 3 = 2\sqrt{x}$$
$$(x - 3)^2 = \left(2\sqrt{x}\right)^2$$
$$x^2 - 6x + 9 = 4x$$
$$x^2 - 10x + 9 = 0$$
$$(x - 9)(x - 1) = 0$$
$$x = 9 \qquad (x = 1 \text{ does not check in the original equation})$$

The number is 9.

47. Substituting $T = 2$:

$$2 = \frac{11}{7}\sqrt{\frac{L}{2}}$$

$$14 = 11\sqrt{\frac{L}{2}}$$

$$\frac{14}{11} = \sqrt{\frac{L}{2}}$$

$$\left(\frac{14}{11}\right)^2 = \left(\sqrt{\frac{L}{2}}\right)^2$$

$$\frac{196}{121} = \frac{L}{2}$$

$$L = \frac{392}{121} \approx 3.24 \text{ feet}$$

49. Reducing the rational expression: $\dfrac{x^2 - x - 6}{x^2 - 9} = \dfrac{(x-3)(x+2)}{(x+3)(x-3)} = \dfrac{x+2}{x+3}$

51. Performing the operations:

$$\frac{x^2 - 25}{x+4} \cdot \frac{2x+8}{x^2 - 9x + 20} = \frac{(x+5)(x-5)}{x+4} \cdot \frac{2(x+4)}{(x-4)(x-5)} = \frac{2(x+5)(x-5)(x+4)}{(x+4)(x-4)(x-5)} = \frac{2(x+5)}{x-4}$$

53. Performing the operations: $\dfrac{x}{x^2 - 16} + \dfrac{4}{x^2 - 16} = \dfrac{x+4}{x^2 - 16} = \dfrac{x+4}{(x+4)(x-4)} = \dfrac{1}{x-4}$

55. Simplifying the complex fraction: $\dfrac{1 - \dfrac{25}{x^2}}{1 - \dfrac{8}{x} + \dfrac{15}{x^2}} \cdot \dfrac{x^2}{x^2} = \dfrac{x^2 - 25}{x^2 - 8x + 15} = \dfrac{(x+5)(x-5)}{(x-5)(x-3)} = \dfrac{x+5}{x-3}$

57. Multiplying each side of the equation by $x^2 - 9 = (x+3)(x-3)$:

$$(x+3)(x-3)\left(\frac{x}{x^2-9} - \frac{3}{x-3}\right) = (x+3)(x-3) \cdot \frac{1}{x+3}$$

$$x - 3(x+3) = x - 3$$

$$x - 3x - 9 = x - 3$$

$$-2x - 9 = x - 3$$

$$-3x = 6$$

$$x = -2$$

Since $x = -2$ checks in the original equation, the solution is $x = -2$.

59. Let t represent the time to fill the pool with both pipes open. The equation is:

$$\tfrac{1}{8} - \tfrac{1}{12} = \tfrac{1}{t}$$

$$24t\left(\tfrac{1}{8} - \tfrac{1}{12}\right) = 24t \cdot \tfrac{1}{t}$$

$$3t - 2t = 24$$

$$t = 24$$

It will take 24 hours to fill the pool with both pipes left open.

61. The variation equation is $y = Kx$. Finding K:

$$8 = K \cdot 12$$

$$K = \tfrac{2}{3}$$

So $y = \tfrac{2}{3}x$. Substituting $x = 36$: $y = \tfrac{2}{3}(36) = 24$

Chapter 8 Review

1. Finding the root: $\sqrt{25} = 5$

3. Finding the root: $\sqrt[3]{-1} = -1$

5. Finding the root: $\sqrt{100x^2y^4} = 10xy^2$

7. Simplifying the expression: $\sqrt{24} = \sqrt{4 \cdot 6} = 2\sqrt{6}$

9. Simplifying the expression: $\sqrt{90x^3y^4} = \sqrt{9x^2y^4 \cdot 10x} = 3xy^2\sqrt{10x}$

11. Simplifying the expression: $3\sqrt{20x^3y} = 3\sqrt{4x^2 \cdot 5xy} = 3 \cdot 2x\sqrt{5xy} = 6x\sqrt{5xy}$

13. Simplifying the expression: $\sqrt{\dfrac{8}{81}} = \dfrac{\sqrt{8}}{\sqrt{81}} = \dfrac{\sqrt{4 \cdot 2}}{9} = \dfrac{2\sqrt{2}}{9}$

15. Simplifying the expression: $\sqrt{\dfrac{49a^2b^2}{16}} = \dfrac{\sqrt{49a^2b^2}}{\sqrt{16}} = \dfrac{7ab}{4}$

17. Simplifying the expression: $\sqrt{\dfrac{40a^2}{121}} = \dfrac{\sqrt{40a^2}}{\sqrt{121}} = \dfrac{\sqrt{4a^2 \cdot 10}}{11} = \dfrac{2a\sqrt{10}}{11}$

19. Simplifying the expression: $\dfrac{3\sqrt{120a^2b^2}}{\sqrt{25}} = \dfrac{3\sqrt{4a^2b^2 \cdot 30}}{5} = \dfrac{3 \cdot 2ab\sqrt{30}}{5} = \dfrac{6ab\sqrt{30}}{5}$

21. Simplifying the radical expression: $\dfrac{2}{\sqrt{7}} \cdot \dfrac{\sqrt{7}}{\sqrt{7}} = \dfrac{2\sqrt{7}}{7}$

23. Simplifying the radical expression: $\sqrt{\dfrac{5}{48}} = \dfrac{\sqrt{5}}{\sqrt{48}} \cdot \dfrac{\sqrt{3}}{\sqrt{3}} = \dfrac{\sqrt{15}}{\sqrt{144}} = \dfrac{\sqrt{15}}{12}$

25. Simplifying the radical expression: $\sqrt{\dfrac{32ab^2}{3}} = \dfrac{\sqrt{32ab^2}}{\sqrt{3}} \cdot \dfrac{\sqrt{3}}{\sqrt{3}} = \dfrac{\sqrt{96ab^2}}{\sqrt{9}} = \dfrac{\sqrt{16b^2 \cdot 6a}}{3} = \dfrac{4b\sqrt{6a}}{3}$

27. Rationalizing the denominator: $\dfrac{3}{\sqrt{3}-4} \cdot \dfrac{\sqrt{3}+4}{\sqrt{3}+4} = \dfrac{3(\sqrt{3}+4)}{3-16} = \dfrac{3\sqrt{3}+12}{-13} = \dfrac{-3\sqrt{3}-12}{13}$

29. Rationalizing the denominator: $\dfrac{3}{\sqrt{5}-\sqrt{2}} \cdot \dfrac{\sqrt{5}+\sqrt{2}}{\sqrt{5}+\sqrt{2}} = \dfrac{3(\sqrt{5}+\sqrt{2})}{5-2} = \dfrac{3(\sqrt{5}+\sqrt{2})}{3} = \sqrt{5}+\sqrt{2}$

31. Rationalizing the denominator: $\dfrac{\sqrt{5}-\sqrt{2}}{\sqrt{5}+\sqrt{2}} \cdot \dfrac{\sqrt{5}-\sqrt{2}}{\sqrt{5}-\sqrt{2}} = \dfrac{5-\sqrt{10}-\sqrt{10}+2}{5-2} = \dfrac{7-2\sqrt{10}}{3}$

33. Combining the expressions: $3\sqrt{5} - 7\sqrt{5} = -4\sqrt{5}$

35. Combining the expressions:
$$-2\sqrt{45} - 5\sqrt{80} + 2\sqrt{20} = -2\sqrt{9 \cdot 5} - 5\sqrt{16 \cdot 5} + 2\sqrt{4 \cdot 5}$$
$$= -2 \cdot 3\sqrt{5} - 5 \cdot 4\sqrt{5} + 2 \cdot 2\sqrt{5}$$
$$= -6\sqrt{5} - 20\sqrt{5} + 4\sqrt{5}$$
$$= -22\sqrt{5}$$

37. Combining the expressions:
$$\sqrt{40a^3b^2} - a\sqrt{90ab^2} = \sqrt{4a^2b^2 \cdot 10a} - a\sqrt{9b^2 \cdot 10a} = 2ab\sqrt{10a} - 3ab\sqrt{10a} = -ab\sqrt{10a}$$

39. Multiplying the expressions: $4\sqrt{2}(\sqrt{3}+\sqrt{5}) = 4\sqrt{6} + 4\sqrt{10}$

41. Multiplying the expressions: $(2\sqrt{5}-4)(\sqrt{5}+3) = 2\sqrt{25} - 4\sqrt{5} + 6\sqrt{5} - 12 = 10 + 2\sqrt{5} - 12 = -2 + 2\sqrt{5}$

43. Solving the equation by squaring:
$$\sqrt{x-3} = 3$$
$$(\sqrt{x-3})^2 = (3)^2$$
$$x - 3 = 9$$
$$x = 12$$
This value checks in the original equation.

45. Solving the equation by squaring:
$$5\sqrt{a} = 20$$
$$\sqrt{a} = 4$$
$$(\sqrt{a})^2 = (4)^2$$
$$a = 16$$
This value checks in the original equation.

47. Solving the equation by squaring:

$$\sqrt{2x+1}+10=8$$
$$\sqrt{2x+1}=-2$$
$$\left(\sqrt{2x+1}\right)^2=(-2)^2$$
$$2x+1=4$$
$$2x=3$$
$$x=\tfrac{3}{2}$$

Since $\sqrt{2\cdot\tfrac{3}{2}+1}+10=\sqrt{3+1}+10=\sqrt{4}+10=2+10=12\neq8$, there is no solution to the equation.

49. Using the Pythagorean Theorem:

$$x^2=\left(\sqrt{2}\right)^2+(1)^2=2+1=3$$
$$x=\sqrt{3}$$

Chapters 1-8 Cumulative Review

1. Simplifying the expression: $\left(\tfrac{4}{5}\right)\left(\tfrac{5}{4}\right)=\tfrac{20}{20}=1$

3. Simplifying the expression: $-\left|-\tfrac{1}{2}\right|=-\tfrac{1}{2}$

5. Simplifying the expression: $\dfrac{4^2-8^2}{(4-8)^2}=\dfrac{16-64}{(-4)^2}=\dfrac{-48}{16}=-3$

7. Simplifying the expression: $4x-7x=(4-7)x=-3x$

9. Simplifying the expression: $\dfrac{\left(b^6\right)^2\left(b^3\right)^4}{\left(b^{10}\right)^3}=\dfrac{b^{12}\cdot b^{12}}{b^{30}}=\dfrac{b^{24}}{b^{30}}=b^{24-30}=b^{-6}=\dfrac{1}{b^6}$

11. Simplifying the expression: $\left(\tfrac{1}{2}y+2\right)\left(\tfrac{1}{2}y-2\right)=\left(\tfrac{1}{2}y\right)^2-(2)^2=\tfrac{1}{4}y^2-4$

13. Simplifying the expression: $\dfrac{\dfrac{1}{a+6}+3}{\dfrac{1}{a+6}+2}\cdot\dfrac{a+6}{a+6}=\dfrac{1+3(a+6)}{1+2(a+6)}=\dfrac{1+3a+18}{1+2a+12}=\dfrac{3a+19}{2a+13}$

15. Simplifying the expression: $\sqrt{\dfrac{90a^2}{169}}=\dfrac{\sqrt{90a^2}}{\sqrt{169}}=\dfrac{\sqrt{9a^2\cdot10}}{13}=\dfrac{3a\sqrt{10}}{13}$

17. Simplifying the expression:

$$\dfrac{7a}{a^2-3a-54}+\dfrac{5}{a-9}=\dfrac{7a}{(a-9)(a+6)}+\dfrac{5}{a-9}\cdot\dfrac{a+6}{a+6}$$
$$=\dfrac{7a}{(a-9)(a+6)}+\dfrac{5a+30}{(a-9)(a+6)}$$
$$=\dfrac{7a+5a+30}{(a-9)(a+6)}$$
$$=\dfrac{12a+30}{(a-9)(a+6)}$$
$$=\dfrac{6(2a+5)}{(a-9)(a+6)}$$

19. Solving the equation:
$$3(5x - 1) = 6(2x + 3) - 21$$
$$15x - 3 = 12x + 18 - 21$$
$$15x - 3 = 12x - 3$$
$$3x - 3 = -3$$
$$3x = 0$$
$$x = 0$$

21. Setting each factor equal to 0 results in $x = 0$, $x = -\frac{2}{3}$, or $x = 4$.

23. Multiplying each side of the equation by $16a$:
$$16a\left(\frac{4}{a}\right) = 16a\left(\frac{a}{16}\right)$$
$$64 = a^2$$
$$a^2 - 64 = 0$$
$$(a + 8)(a - 8) = 0$$
$$a = -8, 8$$
Both values check in the original equation.

25. Adding the two equations:
$$2x = 4$$
$$x = 2$$
Substituting into the first equation:
$$2 + y = 1$$
$$y = -1$$
The solution is $(2, -1)$.

27. Substituting into the first equation:
$$x - (3x - 1) = 5$$
$$x - 3x + 1 = 5$$
$$-2x + 1 = 5$$
$$-2x = 4$$
$$x = -2$$
Substituting into the second equation: $y = 3(-2) - 1 = -6 - 1 = -7$
The solution is $(-2, -7)$. Note this system could also be solved by graphing:

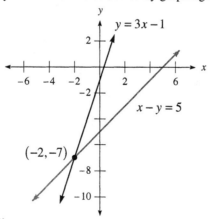

29. Graphing the interval on a number line:

31.   Graphing the line:

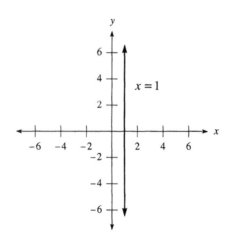

33.   To find the x-intercept, let $y = 0$:
$$0 = -x + 7$$
$$x = 7$$
To find the y-intercept, let $x = 0$: $y = -(0) + 7 = 7$

35.   Using the point-slope formula:
$$y - (-3) = -1(x - 3)$$
$$y + 3 = -x + 3$$
$$y = -x$$

37.   Factoring the polynomial: $r^2 + r - 20 = (r + 5)(r - 4)$

39.   Factoring the polynomial: $x^5 - x^4 - 30x^3 = x^3(x^2 - x - 30) = x^3(x - 6)(x + 5)$

41.   Simplifying the expression: $\dfrac{(6 \times 10^5)(6 \times 10^{-3})}{9 \times 10^{-4}} = \dfrac{36 \times 10^2}{9 \times 10^{-4}} = 4 \times 10^6$

43.   Rationalizing the denominator: $\dfrac{5}{\sqrt{3}} \cdot \dfrac{\sqrt{3}}{\sqrt{3}} = \dfrac{5\sqrt{3}}{\sqrt{9}} = \dfrac{5\sqrt{3}}{3}$

45.   Combining the expressions:
$$5\sqrt{63x^2} - x\sqrt{28} = 5\sqrt{9x^2 \cdot 7} - x\sqrt{4 \cdot 7} = 5 \cdot 3x\sqrt{7} - x \cdot 2\sqrt{7} = 15x\sqrt{7} - 2x\sqrt{7} = 13x\sqrt{7}$$

47.   Reducing the rational expression: $\dfrac{x^2 - 9}{x^4 - 81} = \dfrac{1(x^2 - 9)}{(x^2 + 9)(x^2 - 9)} = \dfrac{1}{x^2 + 9}$

49.   Let x and $2x + 5$ represent the two numbers. The equation is:
$$x + 2x + 5 = 35$$
$$3x + 5 = 35$$
$$3x = 30$$
$$x = 10$$
$$2x + 5 = 25$$
The two numbers are 10 and 25.

Chapter 8 Test

1. Finding the root: $\sqrt{16} = 4$

2. Finding the root. $-\sqrt{36} = -6$

3. The roots are $\sqrt{49} = 7$ and $-\sqrt{49} = -7$.

4. Finding the root: $\sqrt[3]{27} = 3$

5. Finding the root: $\sqrt[3]{-8} = -2$

6. Finding the root: $-\sqrt[4]{81} = -3$

7. Simplifying the expression: $\sqrt{75} = \sqrt{25 \cdot 3} = 5\sqrt{3}$

8. Simplifying the expression: $\sqrt{32} = \sqrt{16 \cdot 2} = 4\sqrt{2}$

9. Simplifying the expression: $\sqrt{\dfrac{2}{3}} = \dfrac{\sqrt{2}}{\sqrt{3}} \cdot \dfrac{\sqrt{3}}{\sqrt{3}} = \dfrac{\sqrt{6}}{\sqrt{9}} = \dfrac{\sqrt{6}}{3}$

10. Simplifying the expression: $\dfrac{1}{\sqrt[3]{4}} \cdot \dfrac{\sqrt[3]{2}}{\sqrt[3]{2}} = \dfrac{\sqrt[3]{2}}{\sqrt[3]{8}} = \dfrac{\sqrt[3]{2}}{2}$

11. Simplifying the expression: $3\sqrt{50x^2} = 3\sqrt{25x^2 \cdot 2} = 3 \cdot 5x\sqrt{2} = 15x\sqrt{2}$

12. Simplifying the expression: $\sqrt{\dfrac{12x^2y^3}{5}} = \dfrac{\sqrt{12x^2y^3}}{\sqrt{5}} \cdot \dfrac{\sqrt{5}}{\sqrt{5}} = \dfrac{\sqrt{60x^2y^3}}{\sqrt{25}} = \dfrac{\sqrt{4x^2y^2 \cdot 15y}}{5} = \dfrac{2xy\sqrt{15y}}{5}$

13. Combining the radicals: $5\sqrt{12} - 2\sqrt{27} = 5\sqrt{4 \cdot 3} - 2\sqrt{9 \cdot 3} = 5 \cdot 2\sqrt{3} - 2 \cdot 3\sqrt{3} = 10\sqrt{3} - 6\sqrt{3} = 4\sqrt{3}$

14. Combining the radicals: $2x\sqrt{18} + 5\sqrt{2x^2} = 2x\sqrt{9 \cdot 2} + 5\sqrt{x^2 \cdot 2} = 2x \cdot 3\sqrt{2} + 5 \cdot x\sqrt{2} = 6x\sqrt{2} + 5x\sqrt{2} = 11x\sqrt{2}$

15. Multiplying the expressions: $\sqrt{3}\left(\sqrt{5} - 2\right) = \sqrt{15} - 2\sqrt{3}$

16. Multiplying the expressions: $\left(\sqrt{5} + 7\right)\left(\sqrt{5} - 8\right) = \sqrt{25} + 7\sqrt{5} - 8\sqrt{5} - 56 = 5 - \sqrt{5} - 56 = -51 - \sqrt{5}$

17. Multiplying the expressions: $\left(\sqrt{x} + 6\right)\left(\sqrt{x} - 6\right) = \left(\sqrt{x}\right)^2 - (6)^2 = x - 36$

18. Multiplying the expressions: $\left(\sqrt{5} - \sqrt{3}\right)^2 = \left(\sqrt{5}\right)^2 - 2\left(\sqrt{5}\right)\left(\sqrt{3}\right) + \left(\sqrt{3}\right)^2 = 5 - 2\sqrt{15} + 3 = 8 - 2\sqrt{15}$

19. Rationalizing the denominator: $\dfrac{\sqrt{7} - \sqrt{3}}{\sqrt{7} + \sqrt{3}} \cdot \dfrac{\sqrt{7} - \sqrt{3}}{\sqrt{7} - \sqrt{3}} = \dfrac{7 - \sqrt{21} - \sqrt{21} + 3}{7 - 3} = \dfrac{10 - 2\sqrt{21}}{4} = \dfrac{2\left(5 - \sqrt{21}\right)}{4} = \dfrac{5 - \sqrt{21}}{2}$

20. Rationalizing the denominator: $\dfrac{\sqrt{x}}{\sqrt{x} + 5} \cdot \dfrac{\sqrt{x} - 5}{\sqrt{x} - 5} = \dfrac{x - 5\sqrt{x}}{x - 25}$

21. Solving the equation by squaring:
$$\sqrt{2x+1} + 2 = 7$$
$$\sqrt{2x+1} = 5$$
$$\left(\sqrt{2x+1}\right)^2 = (5)^2$$
$$2x + 1 = 25$$
$$2x = 24$$
$$x = 12$$
This value checks in the original equation.

22. Solving the equation by squaring:
$$\sqrt{3x+1} + 6 = 2$$
$$\sqrt{3x+1} = -4$$
$$\left(\sqrt{3x+1}\right)^2 = (-4)^2$$
$$3x + 1 = 16$$
$$3x = 15$$
$$x = 5$$
Since $\sqrt{3 \cdot 5 + 1} + 6 = \sqrt{15 + 1} + 6 = \sqrt{16} + 6 = 4 + 6 = 10 \ne 2$, there is no solution to the equation.

23.   Solving the equation by squaring:
$$\sqrt{2x-3} = x-3$$
$$\left(\sqrt{2x-3}\right)^2 = (x-3)^2$$
$$2x-3 = x^2 - 6x + 9$$
$$0 = x^2 - 8x + 12$$
$$0 = (x-6)(x-2)$$
$$x = 2,6$$

Since $x = 2$ does not check in the original equation, the solution is $x = 6$.

24.   Let x represent the number. The equation is:
$$x - 4 = 3\sqrt{x}$$
$$(x-4)^2 = \left(3\sqrt{x}\right)^2$$
$$x^2 - 8x + 16 = 9x$$
$$x^2 - 17x + 16 = 0$$
$$(x-16)(x-1) = 0$$
$$x = 1,16$$

Since $x = 1$ does not check in the original equation, $x = 16$. The number is 16.

25.   Using the Pythagorean Theorem:
$$x^2 = \left(\sqrt{5}\right)^2 + (1)^2 = 5 + 1 = 6$$
$$x = \sqrt{6}$$

Chapter 9
Quadratic Equations

9.1 More Quadratic Equations

1. Solving the equation:
$$x^2 = 9$$
$$x = \pm\sqrt{9}$$
$$x = \pm 3$$

3. Solving the equation:
$$a^2 = 25$$
$$a = \pm\sqrt{25}$$
$$a = \pm 5$$

5. Solving the equation:
$$y^2 = 8$$
$$y = \pm\sqrt{8}$$
$$y = \pm 2\sqrt{2}$$

7. Solving the equation:
$$2x^2 = 100$$
$$x^2 = 50$$
$$x = \pm\sqrt{50}$$
$$x = \pm 5\sqrt{2}$$

9. Solving the equation:
$$3a^2 = 54$$
$$a^2 = 18$$
$$a = \pm\sqrt{18}$$
$$a = \pm 3\sqrt{2}$$

11. Solving the equation:
$$(x+2)^2 = 4$$
$$x+2 = \pm\sqrt{4}$$
$$x+2 = \pm 2$$
$$x = -2 \pm 2$$
$$x = -4, 0$$

13. Solving the equation:
$$(x+1)^2 = 25$$
$$x+1 = \pm\sqrt{25}$$
$$x+1 = \pm 5$$
$$x = -1 \pm 5$$
$$x = -6, 4$$

15. Solving the equation:
$$(a-5)^2 = 75$$
$$a-5 = \pm\sqrt{75}$$
$$a-5 = \pm 5\sqrt{3}$$
$$a = 5 \pm 5\sqrt{3}$$

17. Solving the equation:
$$(y+1)^2 = 50$$
$$y+1 = \pm\sqrt{50}$$
$$y+1 = \pm 5\sqrt{2}$$
$$y = -1 \pm 5\sqrt{2}$$

19. Solving the equation:
$$(2x+1)^2 = 25$$
$$2x+1 = \pm\sqrt{25}$$
$$2x+1 = \pm 5$$
$$2x = -1 \pm 5$$
$$2x = -6, 4$$
$$x = -3, 2$$

21. Solving the equation:
$$(4a-5)^2 = 36$$
$$4a-5 = \pm\sqrt{36}$$
$$4a-5 = \pm 6$$
$$4a = 5 \pm 6$$
$$4a = -1, 11$$
$$a = -\tfrac{1}{4}, \tfrac{11}{4}$$

23. Solving the equation:
$$(3y-1)^2 = 12$$
$$3y-1 = \pm\sqrt{12}$$
$$3y-1 = \pm 2\sqrt{3}$$
$$3y = 1 \pm 2\sqrt{3}$$
$$y = \frac{1 \pm 2\sqrt{3}}{3}$$

25.   Solving the equation:

$$(6x+2)^2 = 27$$
$$6x+2 = \pm\sqrt{27}$$
$$6x+2 = \pm 3\sqrt{3}$$
$$6x = -2 \pm 3\sqrt{3}$$
$$x = \frac{-2 \pm 3\sqrt{3}}{6}$$

27.   Solving the equation:

$$(3x-9)^2 = 27$$
$$3x-9 = \pm\sqrt{27}$$
$$3x-9 = \pm 3\sqrt{3}$$
$$3x = 9 \pm 3\sqrt{3}$$
$$x = 3 \pm \sqrt{3}$$

29.   Solving the equation:

$$(3x+6)^2 = 45$$
$$3x+6 = \pm\sqrt{45}$$
$$3x+6 = \pm 3\sqrt{5}$$
$$3x = -6 \pm 3\sqrt{5}$$
$$x = -2 \pm \sqrt{5}$$

31.   Solving the equation:

$$(2y-4)^2 = 8$$
$$2y-4 = \pm\sqrt{8}$$
$$2y-4 = \pm 2\sqrt{2}$$
$$2y = 4 \pm 2\sqrt{2}$$
$$y = 2 \pm \sqrt{2}$$

33.   Solving the equation:

$$\left(x-\tfrac{2}{3}\right)^2 = \tfrac{25}{9}$$
$$x-\tfrac{2}{3} = \pm\sqrt{\tfrac{25}{9}}$$
$$x-\tfrac{2}{3} = \pm\tfrac{5}{3}$$
$$x = \tfrac{2}{3} \pm \tfrac{5}{3}$$
$$x = -1, \tfrac{7}{3}$$

35.   Solving the equation:

$$\left(x+\tfrac{1}{2}\right)^2 = \tfrac{7}{4}$$
$$x+\tfrac{1}{2} = \pm\sqrt{\tfrac{7}{4}}$$
$$x+\tfrac{1}{2} = \pm\frac{\sqrt{7}}{2}$$
$$x = \frac{-1 \pm \sqrt{7}}{2}$$

37.   Solving the equation:

$$\left(a-\tfrac{4}{5}\right)^2 = \tfrac{12}{25}$$
$$a-\tfrac{4}{5} = \pm\sqrt{\tfrac{12}{25}}$$
$$a-\tfrac{4}{5} = \pm\frac{2\sqrt{3}}{5}$$
$$a = \frac{4 \pm 2\sqrt{3}}{5}$$

39.   Solving the equation:

$$x^2 + 10x + 25 = 7$$
$$(x+5)^2 = 7$$
$$x+5 = \pm\sqrt{7}$$
$$x = -5 \pm \sqrt{7}$$

41.   Solving the equation:

$$x^2 - 2x + 1 = 9$$
$$(x-1)^2 = 9$$
$$x-1 = \pm\sqrt{9}$$
$$x-1 = \pm 3$$
$$x = 1 \pm 3$$
$$x = -2, 4$$

43.   Solving the equation:

$$x^2 + 12x + 36 = 8$$
$$(x+6)^2 = 8$$
$$x+6 = \pm\sqrt{8}$$
$$x+6 = \pm 2\sqrt{2}$$
$$x = -6 \pm 2\sqrt{2}$$

45.   Checking the solution: $(x+1)^2 = \left(-1+5\sqrt{2}+1\right)^2 = \left(5\sqrt{2}\right)^2 = 25 \cdot 2 = 50$

47.   Let x represent the number. The equation is:

$$(x+3)^2 = 16$$
$$x+3 = \pm\sqrt{16}$$
$$x+3 = \pm 4$$
$$x = -3 \pm 4$$
$$x = -7, 1$$

49.   Solving for r:

$$100(1+r)^2 = A$$
$$(1+r)^2 = \frac{A}{100}$$
$$1+r = \sqrt{\frac{A}{100}} \qquad (\text{since } 1+r > 0)$$
$$1+r = \frac{\sqrt{A}}{10}$$
$$r = -1 + \frac{\sqrt{A}}{10}$$

The number is either –7 or 1.

51.   Let x represent the height of the triangle. Draw the figure:

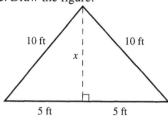

Using the Pythagorean theorem:
$$5^2 + x^2 = 10^2$$
$$25 + x^2 = 100$$
$$x^2 = 75$$
$$x = \sqrt{75}$$
$$x = 5\sqrt{3} \approx 8.66$$
The height is $5\sqrt{3} \approx 8.66$ feet.

53.   Let x represent the height of the triangle. Draw the figure:

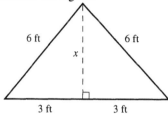

Using the Pythagorean theorem:
$$3^2 + x^2 = 6^2$$
$$9 + x^2 = 36$$
$$x^2 = 27$$
$$x = \sqrt{27}$$
$$x = 3\sqrt{3} \approx 5.2 \text{ feet}$$
No, a person 5 ft 8 in. tall cannot stand up inside the tent.

55.   Let x represent the height of the triangle. Draw the figure:

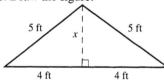

Using the Pythagorean theorem:
$$4^2 + x^2 = 5^2$$
$$16 + x^2 = 25$$
$$x^2 = 9$$
$$x = \sqrt{9} = 3$$
The height is 3 feet.

57.   Substituting $c = 6$, $s = 3.53$, and $d = 192$ in the formula:
$$192 = (3.14)(3.53)(6)\left(\tfrac{1}{2}b\right)^2$$
$$\left(\tfrac{1}{2}b\right)^2 \approx 2.8870$$
$$\tfrac{1}{2}b \approx 1.6991$$
$$b \approx 3.40 \text{ inches}$$

59. Let x represent the distance from the base of the building. Using the Pythagorean theorem:
$$x^2 + 6^2 = 14^2$$
$$x^2 + 36 = 196$$
$$x^2 = 160$$
$$x = \sqrt{160} \approx 12.6$$
The distance is approximately 12.6 feet.

61. Multiplying using the square of binomial formula: $(x-5)^2 = x^2 - 2(x)(5) + (5)^2 = x^2 - 10x + 25$

63. Factoring the polynomial: $x^2 - 12x + 36 = (x-6)(x-6) = (x-6)^2$

65. Factoring the polynomial: $x^2 + 4x + 4 = (x+2)(x+2) = (x+2)^2$

67. Finding the root: $\sqrt[3]{8} = 2$ **69.** Finding the root: $\sqrt[4]{16} = 2$

9.2 Completing the Square

1. The correct term is 9, since: $x^2 + 6x + 9 = (x+3)^2$ **3.** The correct term is 1, since: $x^2 + 2x + 1 = (x+1)^2$

5. The correct term is 16, since: $y^2 - 8y + 16 = (y-4)^2$ **7.** The correct term is 1, since: $y^2 - 2y + 1 = (y-1)^2$

9. The correct term is 64, since: $x^2 + 16x + 64 = (x+8)^2$ **11.** The correct term is $\frac{9}{4}$, since: $a^2 - 3a + \frac{9}{4} = \left(a - \frac{3}{2}\right)^2$

13. The correct term is $\frac{49}{4}$, since: $x^2 - 7x + \frac{49}{4} = \left(x - \frac{7}{2}\right)^2$ **15.** The correct term is $\frac{1}{4}$, since: $y^2 + y + \frac{1}{4} = \left(y + \frac{1}{2}\right)^2$

17. The correct term is $\frac{9}{16}$, since: $x^2 - \frac{3}{2}x + \frac{9}{16} = \left(x - \frac{3}{4}\right)^2$

19. Solve by completing the square:
$$x^2 + 4x = 12$$
$$x^2 + 4x + 4 = 12 + 4$$
$$(x+2)^2 = 16$$
$$x + 2 = \pm 4$$
$$x = -2 \pm 4$$
$$x = -6, 2$$

21. Solve by completing the square:
$$x^2 - 6x = 16$$
$$x^2 - 6x + 9 = 16 + 9$$
$$(x-3)^2 = 25$$
$$x - 3 = \pm 5$$
$$x = 3 \pm 5$$
$$x = -2, 8$$

23. Solve by completing the square:
$$a^2 + 2a = 3$$
$$a^2 + 2a + 1 = 3 + 1$$
$$(a+1)^2 = 4$$
$$a + 1 = \pm 2$$
$$a = -1 \pm 2$$
$$a = -3, 1$$

25. Solve by completing the square:
$$x^2 - 10x = 0$$
$$x^2 - 10x + 25 = 0 + 25$$
$$(x-5)^2 = 25$$
$$x - 5 = \pm 5$$
$$x = 5 \pm 5$$
$$x = 0, 10$$

27. Solve by completing the square:
$$y^2 + 2y - 15 = 0$$
$$y^2 + 2y = 15$$
$$y^2 + 2y + 1 = 15 + 1$$
$$(y+1)^2 = 16$$
$$y + 1 = \pm 4$$
$$y = -1 \pm 4$$
$$y = -5, 3$$

29. Solve by completing the square:
$$x^2 + 4x - 3 = 0$$
$$x^2 + 4x = 3$$
$$x^2 + 4x + 4 = 3 + 4$$
$$(x+2)^2 = 7$$
$$x + 2 = \pm\sqrt{7}$$
$$x = -2 \pm \sqrt{7}$$

31. Solve by completing the square:

$$x^2 - 4x = 4$$
$$x^2 - 4x + 4 = 4 + 4$$
$$(x-2)^2 = 8$$
$$x - 2 = \pm\sqrt{8}$$
$$x - 2 = \pm 2\sqrt{2}$$
$$x = 2 \pm 2\sqrt{2}$$

33. Solve by completing the square:

$$a^2 = 7a + 8$$
$$a^2 - 7a = 8$$
$$a^2 - 7a + \tfrac{49}{4} = 8 + \tfrac{49}{4}$$
$$\left(a - \tfrac{7}{2}\right)^2 = \tfrac{81}{4}$$
$$a - \tfrac{7}{2} = \pm\sqrt{\tfrac{81}{4}}$$
$$a - \tfrac{7}{2} = \pm\tfrac{9}{2}$$
$$a = \tfrac{7}{2} \pm \tfrac{9}{2}$$
$$a = -1, 8$$

35. Solve by completing the square:

$$4x^2 + 8x - 4 = 0$$
$$x^2 + 2x - 1 = 0$$
$$x^2 + 2x = 1$$
$$x^2 + 2x + 1 = 1 + 1$$
$$(x+1)^2 = 2$$
$$x + 1 = \pm\sqrt{2}$$
$$x = -1 \pm \sqrt{2}$$

37. Solve by completing the square:

$$2x^2 + 2x - 4 = 0$$
$$x^2 + x - 2 = 0$$
$$x^2 + x = 2$$
$$x^2 + x + \tfrac{1}{4} = 2 + \tfrac{1}{4}$$
$$\left(x + \tfrac{1}{2}\right)^2 = \tfrac{9}{4}$$
$$x + \tfrac{1}{2} = \pm\sqrt{\tfrac{9}{4}}$$
$$x + \tfrac{1}{2} = \pm\tfrac{3}{2}$$
$$x = -\tfrac{1}{2} \pm \tfrac{3}{2}$$
$$x = -2, 1$$

39. Solve by completing the square:

$$4x^2 + 8x + 1 = 0$$
$$x^2 + 2x + \tfrac{1}{4} = 0$$
$$x^2 + 2x = -\tfrac{1}{4}$$
$$x^2 + 2x + 1 = -\tfrac{1}{4} + 1$$
$$(x+1)^2 = \tfrac{3}{4}$$
$$x + 1 = \pm\sqrt{\tfrac{3}{4}}$$
$$x + 1 = \pm\frac{\sqrt{3}}{2}$$
$$x = -1 \pm \frac{\sqrt{3}}{2}$$
$$x = \frac{-2 \pm \sqrt{3}}{2}$$

41. Solve by completing the square:

$$2x^2 - 2x = 1$$
$$x^2 - x = \tfrac{1}{2}$$
$$x^2 - x + \tfrac{1}{4} = \tfrac{1}{2} + \tfrac{1}{4}$$
$$\left(x - \tfrac{1}{2}\right)^2 = \tfrac{3}{4}$$
$$x - \tfrac{1}{2} = \pm\sqrt{\tfrac{3}{4}}$$
$$x - \tfrac{1}{2} = \pm\frac{\sqrt{3}}{2}$$
$$x = \tfrac{1}{2} \pm \frac{\sqrt{3}}{2}$$
$$x = \frac{1 \pm \sqrt{3}}{2}$$

43. Solve by completing the square:

$$4a^2 - 4a + 1 = 0$$
$$a^2 - a + \tfrac{1}{4} = 0$$
$$\left(a - \tfrac{1}{2}\right)^2 = 0$$
$$a - \tfrac{1}{2} = 0$$
$$a = \tfrac{1}{2}$$

45. Solve by completing the square:

$$3y^2 - 9y = 2$$
$$y^2 - 3y = \tfrac{2}{3}$$
$$y^2 - 3y + \tfrac{9}{4} = \tfrac{2}{3} + \tfrac{9}{4}$$
$$\left(y - \tfrac{3}{2}\right)^2 = \tfrac{8}{12} + \tfrac{27}{12} = \tfrac{35}{12}$$
$$y - \tfrac{3}{2} = \pm\sqrt{\tfrac{35}{12}} \cdot \tfrac{\sqrt{3}}{\sqrt{3}}$$
$$y - \tfrac{3}{2} = \pm\tfrac{\sqrt{105}}{6}$$
$$y = \tfrac{3}{2} \pm \tfrac{\sqrt{105}}{6}$$
$$y = \tfrac{9 \pm \sqrt{105}}{6}$$

47. Using the Pythagorean theorem:

$$(3x)^2 + (4x)^2 = 14^2$$
$$9x^2 + 16x^2 = 196$$
$$25x^2 = 196$$
$$x^2 = \tfrac{196}{25}$$
$$x = \sqrt{\tfrac{196}{25}} = 2.8$$
$$3x = 8.4, \, 4x = 11.2$$

The length (height) of the screen is 8.4 inches and the width of the screen is 11.2 inches.

49. Solving by completing the square:

$$x^2 - 2x - 1 = 0$$
$$x^2 - 2x = 1$$
$$x^2 - 2x + 1 = 1 + 1$$
$$(x - 1)^2 = 2$$
$$x - 1 = \pm\sqrt{2}$$
$$x = 1 \pm \sqrt{2} \approx -0.4, 2.4$$

51. **a.** Adding the two solutions: $\left(-2 + \sqrt{7}\right) + \left(-2 - \sqrt{7}\right) = -4$

b. Multiplying the two solutions: $\left(-2 + \sqrt{7}\right)\left(-2 - \sqrt{7}\right) = (-2)^2 - \left(\sqrt{7}\right)^2 = 4 - 7 = -3$

53. Drawing the diagram:

	x	3
x	x^2	$3x$
3	$3x$	9

55. Drawing the diagram:

	x	1
x	x^2	x
1	x	1

57. Evaluating when $a = 2$: $2a = 2(2) = 4$

59. Evaluating when $a = 2$ and $c = -3$: $4ac = 4(2)(-3) = -24$

61. Evaluating when $a = 2$, $b = 4$, and $c = -3$: $\sqrt{b^2 - 4ac} = \sqrt{(4)^2 - 4(2)(-3)} = \sqrt{16 + 24} = \sqrt{40} = 2\sqrt{10}$

63. Simplifying the radical: $\sqrt{12} = \sqrt{4 \cdot 3} = \sqrt{4}\sqrt{3} = 2\sqrt{3}$

65. Simplifying the radical: $\sqrt{20x^2 y^3} = \sqrt{4x^2 y^2 \cdot 5y} = \sqrt{4x^2 y^2}\sqrt{5y} = 2xy\sqrt{5y}$

67. Simplifying the radical: $\sqrt{\dfrac{81}{25}} = \dfrac{\sqrt{81}}{\sqrt{25}} = \dfrac{9}{5}$

9.3 The Quadratic Formula

1. Using the quadratic formula with $a = 1$, $b = 3$, and $c = 2$:
$$x = \frac{-b \pm \sqrt{b^2 - 4ac}}{2a} = \frac{-3 \pm \sqrt{(3)^2 - 4(1)(2)}}{2(1)} = \frac{-3 \pm \sqrt{9 - 8}}{2} = \frac{-3 \pm 1}{2} = -2, -1$$

3. Using the quadratic formula with $a = 1$, $b = 5$, and $c = 6$:
$$x = \frac{-b \pm \sqrt{b^2 - 4ac}}{2a} = \frac{-5 \pm \sqrt{(5)^2 - 4(1)(6)}}{2(1)} = \frac{-5 \pm \sqrt{25 - 24}}{2} = \frac{-5 \pm 1}{2} = -3, -2$$

5. Using the quadratic formula with $a = 1$, $b = 6$, and $c = 9$:
$$x = \frac{-b \pm \sqrt{b^2 - 4ac}}{2a} = \frac{-6 \pm \sqrt{(6)^2 - 4(1)(9)}}{2(1)} = \frac{-6 \pm \sqrt{36 - 36}}{2} = \frac{-6 \pm 0}{2} = -3$$

7. Using the quadratic formula with $a = 1$, $b = 6$, and $c = 7$:
$$x = \frac{-b \pm \sqrt{b^2 - 4ac}}{2a} = \frac{-6 \pm \sqrt{(6)^2 - 4(1)(7)}}{2(1)} = \frac{-6 \pm \sqrt{36 - 28}}{2} = \frac{-6 \pm \sqrt{8}}{2} = \frac{-6 \pm 2\sqrt{2}}{2} = -3 \pm \sqrt{2}$$

9. Using the quadratic formula with $a = 2$, $b = 5$, and $c = 3$:
$$x = \frac{-b \pm \sqrt{b^2 - 4ac}}{2a} = \frac{-5 \pm \sqrt{(5)^2 - 4(2)(3)}}{2(2)} = \frac{-5 \pm \sqrt{25 - 24}}{4} = \frac{-5 \pm 1}{4} = -\frac{3}{2}, -1$$

11. Using the quadratic formula with $a = 4$, $b = 8$, and $c = 1$:
$$x = \frac{-b \pm \sqrt{b^2 - 4ac}}{2a} = \frac{-8 \pm \sqrt{(8)^2 - 4(4)(1)}}{2(4)} = \frac{-8 \pm \sqrt{64 - 16}}{8} = \frac{-8 \pm \sqrt{48}}{8} = \frac{-8 \pm 4\sqrt{3}}{8} = \frac{-2 \pm \sqrt{3}}{2}$$

13. Using the quadratic formula with $a = 1$, $b = -2$, and $c = 1$:
$$x = \frac{-b \pm \sqrt{b^2 - 4ac}}{2a} = \frac{-(-2) \pm \sqrt{(-2)^2 - 4(1)(1)}}{2(1)} = \frac{2 \pm \sqrt{4 - 4}}{2} = \frac{2 \pm 0}{2} = 1$$

15. First write the equation as $x^2 - 5x - 7 = 0$. Using $a = 1$, $b = -5$, and $c = -7$ in the quadratic formula:
$$x = \frac{-b \pm \sqrt{b^2 - 4ac}}{2a} = \frac{-(-5) \pm \sqrt{(-5)^2 - 4(1)(-7)}}{2(1)} = \frac{5 \pm \sqrt{25 + 28}}{2} = \frac{5 \pm \sqrt{53}}{2}$$

17. Using $a = 6$, $b = -1$, and $c = -2$ in the quadratic formula:
$$x = \frac{-b \pm \sqrt{b^2 - 4ac}}{2a} = \frac{-(-1) \pm \sqrt{(-1)^2 - 4(6)(-2)}}{2(6)} = \frac{1 \pm \sqrt{1 + 48}}{12} = \frac{1 \pm \sqrt{49}}{12} = \frac{1 \pm 7}{12} = -\frac{1}{2}, \frac{2}{3}$$

19. First simplify the equation:
$$(x - 2)(x + 1) = 3$$
$$x^2 - 2x + x - 2 = 3$$
$$x^2 - x - 5 = 0$$
Using $a = 1$, $b = -1$, and $c = -5$ in the quadratic formula:
$$x = \frac{-b \pm \sqrt{b^2 - 4ac}}{2a} = \frac{-(-1) \pm \sqrt{(-1)^2 - 4(1)(-5)}}{2(1)} = \frac{1 \pm \sqrt{1 + 20}}{2} = \frac{1 \pm \sqrt{21}}{2}$$

21. First simplify the equation:
$$(2x-3)(x+2)=1$$
$$2x^2-3x+4x-6=1$$
$$2x^2+x-7=0$$
Using $a=2$, $b=1$, and $c=-7$ in the quadratic formula:
$$x=\frac{-b\pm\sqrt{b^2-4ac}}{2a}=\frac{-1\pm\sqrt{(1)^2-4(2)(-7)}}{2(2)}=\frac{-1\pm\sqrt{1+56}}{4}=\frac{-1\pm\sqrt{57}}{4}$$

23. First write the equation as $2x^2-3x-5=0$. Using $a=2$, $b=-3$, and $c=-5$ in the quadratic formula:
$$x=\frac{-b\pm\sqrt{b^2-4ac}}{2a}=\frac{-(-3)\pm\sqrt{(-3)^2-4(2)(-5)}}{2(2)}=\frac{3\pm\sqrt{9+40}}{4}=\frac{3\pm\sqrt{49}}{4}=\frac{3\pm7}{4}=-1,\frac{5}{2}$$

25. First write the equation as $2x^2+6x-7=0$. Using $a=2$, $b=6$, and $c=-7$ in the quadratic formula:
$$x=\frac{-b\pm\sqrt{b^2-4ac}}{2a}=\frac{-6\pm\sqrt{(6)^2-4(2)(-7)}}{2(2)}=\frac{-6\pm\sqrt{36+56}}{4}=\frac{-6\pm\sqrt{92}}{4}=\frac{-6\pm2\sqrt{23}}{4}=\frac{-3\pm\sqrt{23}}{2}$$

27. First write the equation as $3x^2+4x-2=0$. Using $a=3$, $b=4$, and $c=-2$ in the quadratic formula:
$$x=\frac{-b\pm\sqrt{b^2-4ac}}{2a}=\frac{-4\pm\sqrt{(4)^2-4(3)(-2)}}{2(3)}=\frac{-4\pm\sqrt{16+24}}{6}=\frac{-4\pm\sqrt{40}}{6}=\frac{-4\pm2\sqrt{10}}{6}=\frac{-2\pm\sqrt{10}}{3}$$

29. First write the equation as $2x^2-2x-5=0$. Using $a=2$, $b=-2$, and $c=-5$ in the quadratic formula:
$$x=\frac{-b\pm\sqrt{b^2-4ac}}{2a}=\frac{-(-2)\pm\sqrt{(-2)^2-4(2)(-5)}}{2(2)}=\frac{2\pm\sqrt{4+40}}{4}=\frac{2\pm\sqrt{44}}{4}=\frac{2\pm2\sqrt{11}}{4}=\frac{1\pm\sqrt{11}}{2}$$

31. Factoring out x results in the equation $x(2x^2+3x-4)=0$, so $x=0$ is one solution. The other two solutions are found by using $a=2$, $b=3$, and $c=-4$ in the quadratic formula:
$$x=\frac{-b\pm\sqrt{b^2-4ac}}{2a}=\frac{-3\pm\sqrt{(3)^2-4(2)(-4)}}{2(2)}=\frac{-3\pm\sqrt{9+32}}{4}=\frac{-3\pm\sqrt{41}}{4}$$

33. Using $a=3$, $b=-4$, and $c=0$ in the quadratic formula:
$$x=\frac{-b\pm\sqrt{b^2-4ac}}{2a}=\frac{-(-4)\pm\sqrt{(-4)^2-4(3)(0)}}{2(3)}=\frac{4\pm\sqrt{16-0}}{6}=\frac{4\pm4}{6}=0,\frac{4}{3}$$

35. Multiplying each side of the equation by 6 results in the equation $3x^2-3x-1=0$. Using $a=3$, $b=-3$, and $c=-1$ in the quadratic formula: $x=\dfrac{-b\pm\sqrt{b^2-4ac}}{2a}=\dfrac{-(-3)\pm\sqrt{(-3)^2-4(3)(-1)}}{2(3)}=\dfrac{3\pm\sqrt{9+12}}{6}=\dfrac{3\pm\sqrt{21}}{6}$

37. Solving the equation:
$$56=8+64t-16t^2$$
$$16t^2-64t+48=0$$
$$t^2-4t+3=0$$
$$(t-1)(t-3)=0$$
$$t=1,3$$
The arrow is 56 feet above the ground after 1 second and after 3 seconds.

39. Multiplying the radicals: $(2\sqrt{3})(3\sqrt{5})=6\sqrt{15}$

41. Multiplying the radicals: $(\sqrt{6}+2)(\sqrt{6}-5)=6+2\sqrt{6}-5\sqrt{6}-10=-4-3\sqrt{6}$

43. Multiplying the radicals: $(\sqrt{7}-\sqrt{2})(\sqrt{7}+\sqrt{2})=(\sqrt{7})^2-(\sqrt{2})^2=7-2=5$

45. Rationalizing the denominator: $\dfrac{2}{3+\sqrt{5}}\cdot\dfrac{3-\sqrt{5}}{3-\sqrt{5}}=\dfrac{2(3-\sqrt{5})}{9-5}=\dfrac{2(3-\sqrt{5})}{4}=\dfrac{3-\sqrt{5}}{2}$

9.4 Complex Numbers

1.   Combining the complex numbers: $(3-2i)+3i = 3+(-2i+3i) = 3+i$

3.   Combining the complex numbers: $(6+2i)-10i = 6+(2i-10i) = 6-8i$

5.   Combining the complex numbers: $(11+9i)-9i = 11+(9i-9i) = 11$

7.   Combining the complex numbers: $(3+2i)+(6-i) = (3+6)+(2i-i) = 9+i$

9.   Combining the complex numbers: $(5+7i)-(6+8i) = 5+7i-6-8i = -1-i$

11.   Combining the complex numbers: $(9-i)+(2-i) = 9-i+2-i = 11-2i$

13.   Combining the complex numbers: $(6+i)-4i-(2-i) = 6+i-4i-2+i = 4-2i$

15.   Combining the complex numbers: $(6-11i)+3i+(2+i) = 6-11i+3i+2+i = 8-7i$

17.   Combining the complex numbers: $(2+3i)-(6-2i)+(3-i) = 2+3i-6+2i+3-i = -1+4i$

19.   Multiplying the complex numbers: $3(2-i) = 6-3i$

21.   Multiplying the complex numbers: $2i(8-7i) = 16i-14i^2 = 16i-14(-1) = 14+16i$

23.   Multiplying the complex numbers: $(2+i)(4-i) = 8+4i-2i-i^2 = 8+2i-(-1) = 8+2i+1 = 9+2i$

25.   Multiplying the complex numbers: $(2+i)(3-5i) = 6+3i-10i-5i^2 = 6-7i-5(-1) = 6-7i+5 = 11-7i$

27.   Multiplying the complex numbers: $(3+5i)(3-5i) = (3)^2-(5i)^2 = 9-25i^2 = 9-25(-1) = 9+25 = 34$

29.   Multiplying the complex numbers: $(2+i)(2-i) = (2)^2-(i)^2 = 4-i^2 = 4-(-1) = 4+1 = 5$

31.   Dividing the complex numbers: $\dfrac{2}{3-2i} \cdot \dfrac{3+2i}{3+2i} = \dfrac{2(3+2i)}{9-4i^2} = \dfrac{2(3+2i)}{9+4} = \dfrac{6+4i}{13}$

33.   Dividing the complex numbers: $\dfrac{-3i}{2+3i} \cdot \dfrac{2-3i}{2-3i} = \dfrac{-6i+9i^2}{4-9i^2} = \dfrac{-6i-9}{4+9} = \dfrac{-9-6i}{13}$

35.   Dividing the complex numbers: $\dfrac{6i}{3-i} \cdot \dfrac{3+i}{3+i} = \dfrac{18i+6i^2}{9-i^2} = \dfrac{18i-6}{9+1} = \dfrac{6(-1+3i)}{10} = \dfrac{3(-1+3i)}{5} = \dfrac{-3+9i}{5}$

37.   Dividing the complex numbers: $\dfrac{2+i}{2-i} \cdot \dfrac{2+i}{2+i} = \dfrac{4+2i+2i+i^2}{4-i^2} = \dfrac{4+4i-1}{4+1} = \dfrac{3+4i}{5}$

39.   Dividing the complex numbers:

$$\dfrac{4+5i}{3-6i} \cdot \dfrac{3+6i}{3+6i} = \dfrac{12+15i+24i+30i^2}{9-36i^2} = \dfrac{12+39i-30}{9+36} = \dfrac{-18+39i}{45} = \dfrac{3(-6+13i)}{45} = \dfrac{-6+13i}{15}$$

41.   Multiplying the expressions: $(x+3i)(x-3i) = x^2-(3i)^2 = x^2-9i^2 = x^2-9(-1) = x^2+9$

43.   Simplifying: $\dfrac{1}{i} \cdot \dfrac{i}{i} = \dfrac{i}{i^2} = \dfrac{i}{-1} = -i$

45.   Solving the equation:
$$(x-3)^2 = 25$$
$$x-3 = \pm\sqrt{25}$$
$$x-3 = \pm 5$$
$$x = 3 \pm 5$$
$$x = -2, 8$$

47.   Solving the equation:
$$(2x-6)^2 = 16$$
$$2x-6 = \pm\sqrt{16}$$
$$2x-6 = \pm 4$$
$$2x = 6 \pm 4$$
$$2x = 2, 10$$
$$x = 1, 5$$

49.   Solving the equation:
$$(x+3)^2 = 12$$
$$x+3 = \pm\sqrt{12}$$
$$x+3 = \pm 2\sqrt{3}$$
$$x = -3 \pm 2\sqrt{3}$$

51.   Simplifying the radical expression: $\sqrt{\dfrac{1}{2}} = \dfrac{\sqrt{1}}{\sqrt{2}} \cdot \dfrac{\sqrt{2}}{\sqrt{2}} = \dfrac{\sqrt{2}}{2}$

53.   Simplifying the radical expression: $\sqrt{\dfrac{8x^2y^3}{3}} = \dfrac{\sqrt{8x^2y^3}}{\sqrt{3}} \cdot \dfrac{\sqrt{3}}{\sqrt{3}} = \dfrac{\sqrt{24x^2y^3}}{3} = \dfrac{\sqrt{4x^2y^2 \cdot 6y}}{3} = \dfrac{2xy\sqrt{6y}}{3}$

55. Simplifying the radical expression: $\sqrt[3]{\dfrac{1}{4}} = \dfrac{\sqrt[3]{1}}{\sqrt[3]{4}} \cdot \dfrac{\sqrt[3]{2}}{\sqrt[3]{2}} = \dfrac{\sqrt[3]{2}}{\sqrt[3]{8}} = \dfrac{\sqrt[3]{2}}{2}$

9.5 Complex Solutions to Quadratic Equations

1. Writing as a complex number: $\sqrt{-16} = \sqrt{16(-1)} = \sqrt{16}\sqrt{-1} = 4i$

3. Writing as a complex number: $\sqrt{-49} = \sqrt{49(-1)} = \sqrt{49}\sqrt{-1} = 7i$

5. Writing as a complex number: $\sqrt{-6} = \sqrt{6(-1)} = \sqrt{6}\sqrt{-1} = i\sqrt{6}$

7. Writing as a complex number: $\sqrt{-11} = \sqrt{11(-1)} = \sqrt{11}\sqrt{-1} = i\sqrt{11}$

9. Writing as a complex number: $\sqrt{-32} = \sqrt{-16\cdot 2} = \sqrt{-16}\sqrt{2} = 4i\sqrt{2}$

11. Writing as a complex number: $\sqrt{-50} = \sqrt{-25\cdot 2} = \sqrt{-25}\sqrt{2} = 5i\sqrt{2}$

13. Writing as a complex number: $\sqrt{-8} = \sqrt{-4\cdot 2} = \sqrt{-4}\sqrt{2} = 2i\sqrt{2}$

15. Writing as a complex number: $\sqrt{-48} = \sqrt{-16\cdot 3} = \sqrt{-16}\sqrt{3} = 4i\sqrt{3}$

17. First write the equation as $x^2 - 2x + 2 = 0$. Using $a = 1$, $b = -2$, and $c = 2$ in the quadratic formula:
$$x = \frac{-b \pm \sqrt{b^2 - 4ac}}{2a} = \frac{-(-2) \pm \sqrt{(-2)^2 - 4(1)(2)}}{2(1)} = \frac{2 \pm \sqrt{4-8}}{2} = \frac{2 \pm \sqrt{-4}}{2} = \frac{2 \pm 2i}{2} = 1 \pm i$$

19. Solving the equation:
$$x^2 - 4x = -4$$
$$x^2 - 4x + 4 = 0$$
$$(x-2)^2 = 0$$
$$x - 2 = 0$$
$$x = 2$$

21. Solving the equation:
$$2x^2 + 5x = 12$$
$$2x^2 + 5x - 12 = 0$$
$$(2x-3)(x+4) = 0$$
$$x = \tfrac{3}{2}, -4$$

23. Solving the equation:
$$(x-2)^2 = -4$$
$$x - 2 = \pm\sqrt{-4}$$
$$x - 2 = \pm 2i$$
$$x = 2 \pm 2i$$

25. Solving the equation:
$$\left(x + \tfrac{1}{2}\right)^2 = -\tfrac{9}{4}$$
$$x + \tfrac{1}{2} = \pm\sqrt{-\tfrac{9}{4}}$$
$$x + \tfrac{1}{2} = \pm\tfrac{3}{2}i$$
$$x = \frac{-1 \pm 3i}{2}$$

27. Solving the equation:
$$\left(x - \tfrac{1}{2}\right)^2 = -\tfrac{27}{36}$$
$$x - \tfrac{1}{2} = \pm\sqrt{-\tfrac{27}{36}}$$
$$x - \tfrac{1}{2} = \pm\frac{3i\sqrt{3}}{6}$$
$$x - \tfrac{1}{2} = \pm\frac{i\sqrt{3}}{2}$$
$$x = \frac{1 \pm i\sqrt{3}}{2}$$

29. Using $a = 1$, $b = 1$, and $c = 1$ in the quadratic formula:
$$x = \frac{-b \pm \sqrt{b^2 - 4ac}}{2a} = \frac{-1 \pm \sqrt{(1)^2 - 4(1)(1)}}{2(1)} = \frac{-1 \pm \sqrt{1-4}}{2} = \frac{-1 \pm \sqrt{-3}}{2} = \frac{-1 \pm i\sqrt{3}}{2}$$

31. Solving the equation:
$$x^2 - 5x + 6 = 0$$
$$(x-2)(x-3) = 0$$
$$x = 2, 3$$

33.   First multiply by 6 to clear the equation of fractions:
$$6\left(\tfrac{1}{2}x^2 + \tfrac{1}{3}x + \tfrac{1}{6}\right) = 6(0)$$
$$3x^2 + 2x + 1 = 0$$
Using $a = 3$, $b = 2$, and $c = 1$ in the quadratic formula:
$$x = \frac{-b \pm \sqrt{b^2 - 4ac}}{2a} = \frac{-2 \pm \sqrt{(2)^2 - 4(3)(1)}}{2(3)} = \frac{-2 \pm \sqrt{4 - 12}}{6} = \frac{-2 \pm \sqrt{-8}}{6} = \frac{-2 \pm 2i\sqrt{2}}{6} = \frac{-1 \pm i\sqrt{2}}{3}$$

35.   First multiply by 6 to clear the equation of fractions:
$$6\left(\tfrac{1}{3}x^2\right) = 6\left(-\tfrac{1}{2}x + \tfrac{1}{3}\right)$$
$$2x^2 = -3x + 2$$
$$2x^2 + 3x - 2 = 0$$
$$(2x - 1)(x + 2) = 0$$
$$x = \tfrac{1}{2}, -2$$

37.   Solving the equation:
$$(x + 2)(x - 3) = 5$$
$$x^2 - x - 6 = 5$$
$$x^2 - x - 11 = 0$$
Using $a = 1$, $b = -1$, and $c = -11$ in the quadratic formula:
$$x = \frac{-b \pm \sqrt{b^2 - 4ac}}{2a} = \frac{-(-1) \pm \sqrt{(-1)^2 - 4(1)(-11)}}{2(1)} = \frac{1 \pm \sqrt{1 + 44}}{2} = \frac{1 \pm \sqrt{45}}{2} = \frac{1 \pm 3\sqrt{5}}{2}$$

39.   Solving the equation:
$$(x - 5)(x - 3) = -10$$
$$x^2 - 8x + 15 = -10$$
$$x^2 - 8x + 25 = 0$$
Using $a = 1$, $b = -8$, and $c = 25$ in the quadratic formula:
$$x = \frac{-b \pm \sqrt{b^2 - 4ac}}{2a} = \frac{-(-8) \pm \sqrt{(-8)^2 - 4(1)(25)}}{2(1)} = \frac{8 \pm \sqrt{64 - 100}}{2} = \frac{8 \pm \sqrt{-36}}{2} = \frac{8 \pm 6i}{2} = 4 \pm 3i$$

41.   Solving the equation:
$$(2x - 2)(x - 3) = 9$$
$$2x^2 - 8x + 6 = 9$$
$$2x^2 - 8x - 3 = 0$$
Using $a = 2$, $b = -8$, and $c = -3$ in the quadratic formula:
$$x = \frac{-b \pm \sqrt{b^2 - 4ac}}{2a} = \frac{-(-8) \pm \sqrt{(-8)^2 - 4(2)(-3)}}{2(2)} = \frac{8 \pm \sqrt{64 + 24}}{4} = \frac{8 \pm \sqrt{88}}{4} = \frac{8 \pm 2\sqrt{22}}{4} = \frac{4 \pm \sqrt{22}}{2}$$

43.   Using $a = 1$, $b = -4$, and $c = 5$ in the quadratic formula:
$$x = \frac{-b \pm \sqrt{b^2 - 4ac}}{2a} = \frac{-(-4) \pm \sqrt{(-4)^2 - 4(1)(5)}}{2(1)} = \frac{4 \pm \sqrt{16 - 20}}{2} = \frac{4 \pm \sqrt{-4}}{2} = \frac{4 \pm 2i}{2} = 2 \pm i$$
Since the solutions are non-real complex numbers, the graph cannot cross the x-axis.

45.   Substituting $x = 2 + 2i$ into the equation:
$$x^2 - 4x + 8 = (2 + 2i)^2 - 4(2 + 2i) + 8 = 4 + 8i + 4i^2 - 8 - 8i + 8 = 4 + 8i - 4 - 8 - 8i + 8 = 0$$
Yes, $x = 2 + 2i$ is a solution to the equation.

47.   The other solution is $3 - 7i$.

49.   Graphing the line:

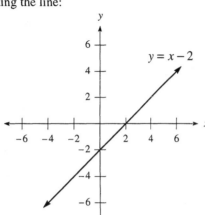

51.   Graphing the line:

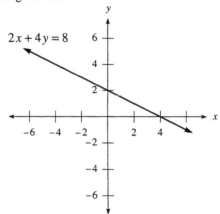

53.   Simplifying the radical expression: $3\sqrt{50} + 2\sqrt{32} = 3\sqrt{25 \cdot 2} + 2\sqrt{16 \cdot 2} = 3 \cdot 5\sqrt{2} + 2 \cdot 4\sqrt{2} = 15\sqrt{2} + 8\sqrt{2} = 23\sqrt{2}$

55.   Simplifying the radical expression: $\sqrt{24} - \sqrt{54} - \sqrt{150} = \sqrt{4 \cdot 6} - \sqrt{9 \cdot 6} - \sqrt{25 \cdot 6} = 2\sqrt{6} - 3\sqrt{6} - 5\sqrt{6} = -6\sqrt{6}$

57.   Simplifying the radical expression:

$$2\sqrt{27x^2} - x\sqrt{48} = 2\sqrt{9x^2 \cdot 3} - x\sqrt{16 \cdot 3} = 2 \cdot 3x\sqrt{3} - x \cdot 4\sqrt{3} = 6x\sqrt{3} - 4x\sqrt{3} = 2x\sqrt{3}$$

9.6 Graphing Parabolas

1.   Graphing the parabola:

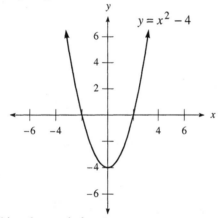

3.   Graphing the parabola:

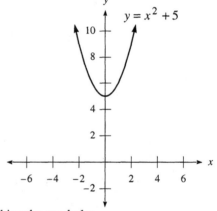

5.   Graphing the parabola:

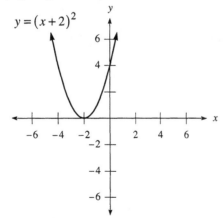

7.   Graphing the parabola:

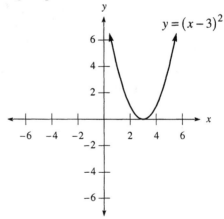

9.   Graphing the parabola:

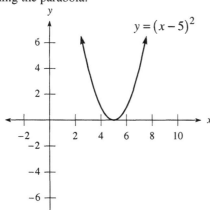

$y = (x - 5)^2$

11.   Graphing the parabola:

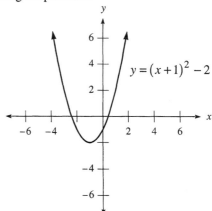

$y = (x + 1)^2 - 2$

13.   Graphing the parabola:

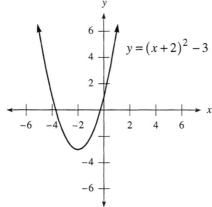

$y = (x + 2)^2 - 3$

15.   Graphing the parabola:

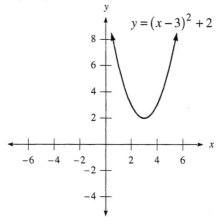

$y = (x - 3)^2 + 2$

17.   Begin by completing the square: $y = x^2 + 6x + 5 = \left(x^2 + 6x + 9\right) + 5 - 9 = (x + 3)^2 - 4$

Graphing the parabola:

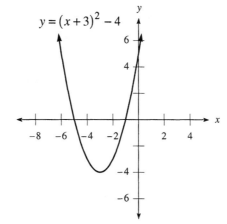

$y = (x + 3)^2 - 4$

19.   Begin by completing the square: $y = x^2 - 2x - 3 = \left(x^2 - 2x + 1\right) - 3 - 1 = (x-1)^2 - 4$

Graphing the parabola:

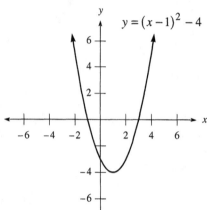

21.   Begin by making a table of values:

x	y
-3	-5
-2	0
-1	3
0	4
1	3
2	0
3	-5

Graphing the parabola:

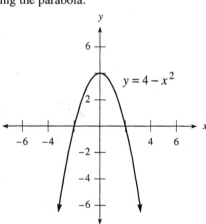

23.   Begin by making a table of values:

x	y
-2	-5
-1	-2
0	-1
1	-2
2	-5

Graphing the parabola:

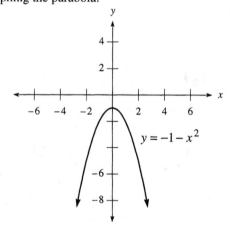

25.   Graphing the line and the parabola:

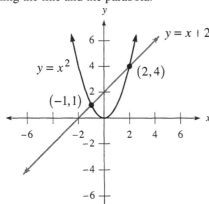

27.   Graphing the two parabolas:

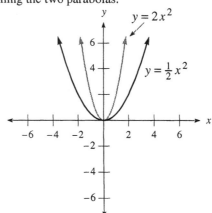

The intersection points are $(-1,1)$ and $(2,4)$.

29.   **a.**   Graphing the orbit:

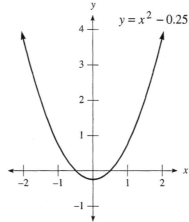

b.   The distance is 0.25 million miles, or 250,000 miles.

c.   The comet crosses the x-axis when $x = 0.5$, which corresponds to 0.5 million miles, or 500,000 miles.

31.   Finding the root: $\sqrt{49} = 7$

33.   Simplifying the radical: $\sqrt{50} = \sqrt{25 \cdot 2} = 5\sqrt{2}$

35.   Simplifying the radical: $\sqrt{\dfrac{2}{5}} = \dfrac{\sqrt{2}}{\sqrt{5}} \cdot \dfrac{\sqrt{5}}{\sqrt{5}} = \dfrac{\sqrt{10}}{5}$

37.   Performing the operations: $3\sqrt{12} + 5\sqrt{27} = 3\sqrt{4 \cdot 3} + 5\sqrt{9 \cdot 3} = 3 \cdot 2\sqrt{3} + 5 \cdot 3\sqrt{3} = 6\sqrt{3} + 15\sqrt{3} = 21\sqrt{3}$

39.   Performing the operations: $\left(\sqrt{6} + 2\right)\left(\sqrt{6} - 5\right) = \sqrt{36} + 2\sqrt{6} - 5\sqrt{6} - 10 = 6 - 3\sqrt{6} - 10 = -4 - 3\sqrt{6}$

41.   Rationalizing the denominator: $\dfrac{8}{\sqrt{5} - \sqrt{3}} \cdot \dfrac{\sqrt{5} + \sqrt{3}}{\sqrt{5} + \sqrt{3}} = \dfrac{8\left(\sqrt{5} + \sqrt{3}\right)}{5 - 3} = \dfrac{8\left(\sqrt{5} + \sqrt{3}\right)}{2} = 4\left(\sqrt{5} + \sqrt{3}\right) = 4\sqrt{5} + 4\sqrt{3}$

43.   Solving the equation:
$$\sqrt{2x - 5} = 3$$
$$\left(\sqrt{2x - 5}\right)^2 = 3^2$$
$$2x - 5 = 9$$
$$2x = 14$$
$$x = 7$$

This solution checks in the original equation.

Chapter 9 Review

1. Solving the quadratic equation:
$$a^2 = 32$$
$$a = \pm\sqrt{32}$$
$$a = \pm 4\sqrt{2}$$

3. Solving the quadratic equation:
$$2x^2 = 32$$
$$x^2 = 16$$
$$x = \pm\sqrt{16}$$
$$x = \pm 4$$

5. Solving the quadratic equation:
$$(x-2)^2 = 81$$
$$x-2 = \pm\sqrt{81}$$
$$x-2 = \pm 9$$
$$x = 2 \pm 9$$
$$x = -7, 11$$

7. Solving the quadratic equation:
$$(2x+5)^2 = 32$$
$$2x+5 = \pm\sqrt{32}$$
$$2x+5 = \pm 4\sqrt{2}$$
$$2x = -5 \pm 4\sqrt{2}$$
$$x = \frac{-5 \pm 4\sqrt{2}}{2}$$

9. Solving the quadratic equation:
$$\left(x-\tfrac{2}{3}\right)^2 = -\tfrac{25}{9}$$
$$x-\tfrac{2}{3} = \pm\sqrt{-\tfrac{25}{9}}$$
$$x-\tfrac{2}{3} = \pm\frac{5i}{3}$$
$$x = \tfrac{2}{3} \pm \tfrac{5}{3}i$$

11. Solving by completing the square:
$$x^2 - 8x = 4$$
$$x^2 - 8x + 16 = 4 + 16$$
$$(x-4)^2 = 20$$
$$x-4 = \pm\sqrt{20}$$
$$x-4 = \pm 2\sqrt{5}$$
$$x = 4 \pm 2\sqrt{5}$$

13. Solving by completing the square:
$$x^2 + 4x + 3 = 0$$
$$x^2 + 4x = -3$$
$$x^2 + 4x + 4 = -3 + 4$$
$$(x+2)^2 = 1$$
$$x+2 = \pm\sqrt{1}$$
$$x+2 = \pm 1$$
$$x = -2 \pm 1$$
$$x = -3, -1$$

15. Solving by completing the square:
$$a^2 = 5a + 6$$
$$a^2 - 5a = 6$$
$$a^2 - 5a + \tfrac{25}{4} = 6 + \tfrac{25}{4}$$
$$\left(a-\tfrac{5}{2}\right)^2 = \tfrac{49}{4}$$
$$a-\tfrac{5}{2} = \pm\sqrt{\tfrac{49}{4}}$$
$$a-\tfrac{5}{2} = \pm\tfrac{7}{2}$$
$$a = \tfrac{5}{2} \pm \tfrac{7}{2}$$
$$a = -1, 6$$

17. Solving by completing the square:
$$3x^2 - 6x - 2 = 0$$
$$x^2 - 2x - \tfrac{2}{3} = 0$$
$$x^2 - 2x = \tfrac{2}{3}$$
$$x^2 - 2x + 1 = \tfrac{2}{3} + 1$$
$$(x-1)^2 = \tfrac{5}{3}$$
$$x-1 = \pm\sqrt{\tfrac{5}{3}}$$
$$x-1 = \pm\frac{\sqrt{5}}{\sqrt{3}} \cdot \frac{\sqrt{3}}{\sqrt{3}}$$
$$x-1 = \pm\frac{\sqrt{15}}{3}$$
$$x = 1 \pm \frac{\sqrt{15}}{3}$$
$$x = \frac{3 \pm \sqrt{15}}{3}$$

19. Using $a = 1$, $b = -8$, and $c = 16$ in the quadratic formula:

$$x = \frac{-b \pm \sqrt{b^2 - 4ac}}{2a} = \frac{-(-8) \pm \sqrt{(-8)^2 - 4(1)(16)}}{2(1)} = \frac{8 \pm \sqrt{64 - 64}}{2} = \frac{8 \pm 0}{2} = 4$$

21. First write the equation as $2x^2 + 8x - 5 = 0$. Using $a = 2$, $b = 8$, and $c = -5$ in the quadratic formula:

$$x = \frac{-b \pm \sqrt{b^2 - 4ac}}{2a} = \frac{-8 \pm \sqrt{(8)^2 - 4(2)(-5)}}{2(2)} = \frac{-8 \pm \sqrt{64 + 40}}{4} = \frac{-8 \pm \sqrt{104}}{4} = \frac{-8 \pm 2\sqrt{26}}{4} = \frac{-4 \pm \sqrt{26}}{2}$$

23. First multiply by 10 to clear the equation of fractions:

$$10\left(\tfrac{1}{5}x^2 - \tfrac{1}{2}x\right) = 10\left(\tfrac{3}{10}\right)$$
$$2x^2 - 5x = 3$$
$$2x^2 - 5x - 3 = 0$$

Using $a = 2$, $b = -5$, and $c = -3$ in the quadratic formula:

$$x = \frac{-b \pm \sqrt{b^2 - 4ac}}{2a} = \frac{-(-5) \pm \sqrt{(-5)^2 - 4(2)(-3)}}{2(2)} = \frac{5 \pm \sqrt{25 + 24}}{4} = \frac{5 \pm \sqrt{49}}{4} = \frac{5 \pm 7}{4} = -\tfrac{1}{2}, 3$$

25. Combining the complex numbers: $(4 - 3i) + 5i = 4 + 2i$

27. Combining the complex numbers: $(5 + 6i) + (5 - i) = 10 + 5i$

29. Combining the complex numbers: $(3 - 2i) - (3 - i) = 3 - 2i - 3 + i = -i$

31. Combining the complex numbers: $(3 + i) - 5i - (4 - i) = 3 + i - 5i - 4 + i = -1 - 3i$

33. Multiplying the complex numbers: $2(3 - i) = 6 - 2i$

35. Multiplying the complex numbers: $4i(6 - 5i) = 24i - 20i^2 = 24i - 20(-1) = 20 + 24i$

37. Multiplying the complex numbers: $(3 - 4i)(5 + i) = 15 - 20i + 3i - 4i^2 = 15 - 17i - 4(-1) = 15 - 17i + 4 = 19 - 17i$

39. Multiplying the complex numbers: $(4 + i)(4 - i) = 16 - i^2 = 16 - (-1) = 16 + 1 = 17$

41. Dividing the complex numbers: $\dfrac{i}{3 + i} \cdot \dfrac{3 - i}{3 - i} = \dfrac{3i - i^2}{9 - i^2} = \dfrac{3i + 1}{9 + 1} = \dfrac{1 + 3i}{10}$

43. Dividing the complex numbers: $\dfrac{5}{2 + 5i} \cdot \dfrac{2 - 5i}{2 - 5i} = \dfrac{10 - 25i}{4 - 25i^2} = \dfrac{10 - 25i}{4 + 25} = \dfrac{10 - 25i}{29}$

45. Dividing the complex numbers: $\dfrac{-3i}{3 - 2i} \cdot \dfrac{3 + 2i}{3 + 2i} = \dfrac{-9i - 6i^2}{9 - 4i^2} = \dfrac{-9i + 6}{9 + 4} = \dfrac{6 - 9i}{13}$

47. Dividing the complex numbers: $\dfrac{4 - 5i}{4 + 5i} \cdot \dfrac{4 - 5i}{4 - 5i} = \dfrac{16 - 20i - 20i + 25i^2}{16 - 25i^2} = \dfrac{16 - 40i - 25}{16 + 25} = \dfrac{-9 - 40i}{41}$

49. Writing as a complex number: $\sqrt{-36} = 6i$

51. Writing as a complex number: $\sqrt{-17} = i\sqrt{17}$

53. Writing as a complex number: $\sqrt{-40} = \sqrt{-4 \cdot 10} = 2i\sqrt{10}$

55. Writing as a complex number: $\sqrt{-200} = \sqrt{-100 \cdot 2} = 10i\sqrt{2}$

57. Graphing the parabola:

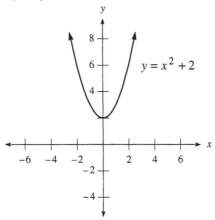

59. Graphing the parabola:

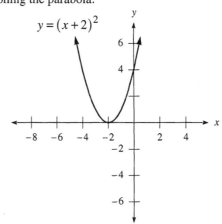

61. Begin by completing the square: $y = x^2 + 4x + 7 = \left(x^2 + 4x + 4\right) + 7 - 4 = (x+2)^2 + 3$

Graphing the parabola:

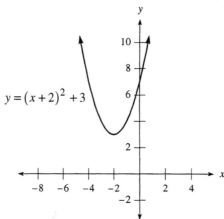

$$y = (x+2)^2 + 3$$

Chapters 1-9 Cumulative Review

1. Simplifying the expression: $10 - 8 - 11 = 10 + (-8) + (-11) = -9$

3. Simplifying the expression: $-\frac{4}{5} \div \frac{8}{15} = -\frac{4}{5} \cdot \frac{15}{8} = -\frac{60}{40} = -\frac{3}{2}$

5. Simplifying the expression: $\dfrac{\left(3x^5\right)\left(20x^3\right)}{15x^{10}} = \dfrac{60x^8}{15x^{10}} = 4x^{-2} = \dfrac{4}{x^2}$

7. Simplifying the expression: $\sqrt{81} = 9$

9. Simplifying the expression: $\sqrt{\dfrac{200}{81}} = \dfrac{\sqrt{200}}{\sqrt{81}} = \dfrac{\sqrt{100 \cdot 2}}{9} = \dfrac{10\sqrt{2}}{9}$

11. Simplifying the expression: $(3 + 3i) - 7i - (2 + 2i) = 3 + 3i - 7i - 2 - 2i = 1 - 6i$

13. Multiplying using the column method:

$$
\begin{array}{r}
a^2 - 6a + 7 \\
a - 2 \\
\hline
a^3 - 6a^2 + 7a \\
-2a^2 + 12a - 14 \\
\hline
a^3 - 8a^2 + 19a - 14
\end{array}
$$

15. Solving the equation:

$$
\begin{aligned}
7 - 4(3x + 4) &= -9x \\
7 - 12x - 16 &= -9x \\
-12x - 9 &= -9x \\
-9 &= 3x \\
x &= -3
\end{aligned}
$$

17. Solving the equation:

$$
\begin{aligned}
5x^2 &= -15x \\
5x^2 + 15x &= 0 \\
5x(x + 3) &= 0 \\
x &= 0, -3
\end{aligned}
$$

19. Solving the equation:

$$
\begin{aligned}
\sqrt{6x - 2} &= 3x - 5 \\
\left(\sqrt{6x - 2}\right)^2 &= (3x - 5)^2 \\
6x - 2 &= 9x^2 - 30x + 25 \\
0 &= 9x^2 - 36x + 27 \\
0 &= x^2 - 4x + 3 \\
0 &= (x - 3)(x - 1) \\
x &= 1, 3
\end{aligned}
$$

Note that $\sqrt{6 \cdot 1 - 2} = \sqrt{6 - 2} = \sqrt{4} = 2$, while $3 \cdot 1 - 5 = 3 - 5 = -2$, so $x = 1$ does not check. The solution is $x = 3$, which checks.

21. First multiply by 6 to clear the equation of fractions:

$$6\left(\tfrac{1}{2}x^2 - \tfrac{1}{3}x\right) = 6\left(-\tfrac{1}{6}\right)$$

$$3x^2 - 2x = -1$$

$$3x^2 - 2x + 1 = 0$$

Using $a = 3$, $b = -2$, and $c = 1$ in the quadratic formula:

$$x = \frac{-b \pm \sqrt{b^2 - 4ac}}{2a} = \frac{-(-2) \pm \sqrt{(-2)^2 - 4(3)(1)}}{2(3)} = \frac{2 \pm \sqrt{4 - 12}}{6} = \frac{2 \pm \sqrt{-8}}{6} = \frac{2 \pm 2i\sqrt{2}}{6} = \frac{1 \pm i\sqrt{2}}{3}$$

23. Solving the inequality:

$$-5 \le 2x - 1 \le 7$$

$$-4 \le 2x \le 8$$

$$-2 \le x \le 4$$

Graphing the solution set:

25. Graphing the line:

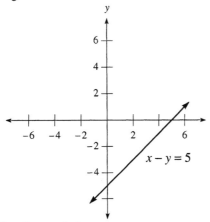

$$x - y = 5$$

27. Graphing the line:

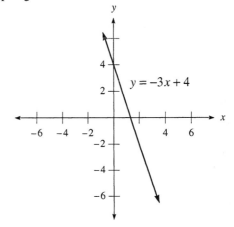

$$y = -3x + 4$$

29. Graphing the parabola:

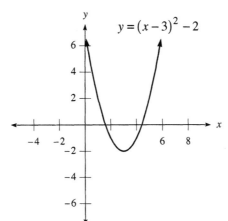

$$y = (x - 3)^2 - 2$$

31. Substituting into the second equation:

$$2(y - 3) + 3y = 4$$

$$2y - 6 + 3y = 4$$

$$5y - 6 = 4$$

$$5y = 10$$

$$y = 2$$

$$x = 2 - 3 = -1$$

The solution is $(-1, 2)$.

33.   Factoring the polynomial: $x^2 - 5x - 24 = (x-8)(x+3)$

35.   Factoring the polynomial: $25 - y^2 = (5+y)(5-y)$

37.   Finding the slope: $m = \dfrac{8-(-2)}{1-3} = \dfrac{8+2}{-2} = \dfrac{10}{-2} = -5$

39.   Subtracting: $6-(-2) = 6+2 = 8$

41.   Dividing by the monomial: $\dfrac{15a^3b - 10a^2b^2 - 20ab^3}{5ab} = \dfrac{15a^3b}{5ab} - \dfrac{10a^2b^2}{5ab} - \dfrac{20ab^3}{5ab} = 3a^2 - 2ab - 4b^2$

43.   Simplifying the rational expression: $\dfrac{x^2 + 5x - 24}{x+8} = \dfrac{(x+8)(x-3)}{x+8} = x-3$

45.   Simplifying the rational expression:
$$\dfrac{-1}{x^2-4} - \dfrac{-2}{x^2-4x-12} = \dfrac{-1}{(x+2)(x-2)} + \dfrac{2}{(x-6)(x+2)}$$
$$= \dfrac{-1(x-6)}{(x+2)(x-2)(x-6)} + \dfrac{2(x-2)}{(x+2)(x-2)(x-6)}$$
$$= \dfrac{-x+6+2x-4}{(x+2)(x-2)(x-6)}$$
$$= \dfrac{x+2}{(x+2)(x-2)(x-6)}$$
$$= \dfrac{1}{(x-2)(x-6)}$$

47.   Combining the radicals: $5\sqrt{200} + 9\sqrt{50} = 5\sqrt{100\cdot 2} + 9\sqrt{25\cdot 2} = 5\cdot 10\sqrt{2} + 9\cdot 5\sqrt{2} = 50\sqrt{2} + 45\sqrt{2} = 95\sqrt{2}$

49.   Solving for y:
$$3x - 8y = 24$$
$$-8y = -3x + 24$$
$$y = \tfrac{3}{8}x - 3$$

Chapter 9 Test

1.   Solving the equation:
$$x^2 - 7x - 8 = 0$$
$$(x-8)(x+1) = 0$$
$$x = -1, 8$$

2.   Solving the equation:
$$(x-3)^2 = 12$$
$$x - 3 = \pm\sqrt{12}$$
$$x - 3 = \pm 2\sqrt{3}$$
$$x = 3 \pm 2\sqrt{3}$$

3.   Solving the equation:
$$\left(x - \tfrac{5}{2}\right)^2 = -\tfrac{75}{4}$$
$$x - \tfrac{5}{2} = \pm\sqrt{-\tfrac{75}{4}}$$
$$x - \tfrac{5}{2} = \pm\dfrac{5i\sqrt{3}}{2}$$
$$x = \dfrac{5 \pm 5i\sqrt{3}}{2}$$

4.   First multiply by 6 to clear the equation of fractions:
$$6\left(\tfrac{1}{3}x^2\right) = 6\left(\tfrac{1}{2}x - \tfrac{5}{6}\right)$$
$$2x^2 = 3x - 5$$
$$2x^2 - 3x + 5 = 0$$
Using $a = 2$, $b = -3$, and $c = 5$ in the quadratic formula:
$$x = \dfrac{-b \pm \sqrt{b^2 - 4ac}}{2a} = \dfrac{-(-3) \pm \sqrt{(-3)^2 - 4(2)(5)}}{2(2)} = \dfrac{3 \pm \sqrt{9-40}}{4} = \dfrac{3 \pm \sqrt{-31}}{4} = \dfrac{3 \pm i\sqrt{31}}{4}$$

5. Solving the equation:
$$3x^2 = -2x + 1$$
$$3x^2 + 2x - 1 = 0$$
$$(3x - 1)(x + 1) = 0$$
$$x = \tfrac{1}{3}, -1$$

6. Simplifying the equation:
$$(x + 2)(x - 1) = 6$$
$$x^2 + x - 2 = 6$$
$$x^2 + x - 8 = 0$$
Using $a = 1$, $b = 1$, and $c = -8$ in the quadratic formula:
$$x = \frac{-b \pm \sqrt{b^2 - 4ac}}{2a} = \frac{-1 \pm \sqrt{(1)^2 - 4(1)(-8)}}{2(1)} = \frac{-1 \pm \sqrt{1 + 32}}{2} = \frac{-1 \pm \sqrt{33}}{2}$$

7. Solving the equation:
$$9x^2 + 12x + 4 = 0$$
$$(3x + 2)^2 = 0$$
$$3x + 2 = 0$$
$$3x = -2$$
$$x = -\tfrac{2}{3}$$

8. Solving by completing the square:
$$x^2 - 6x - 6 = 0$$
$$x^2 - 6x = 6$$
$$x^2 - 6x + 9 = 6 + 9$$
$$(x - 3)^2 = 15$$
$$x - 3 = \pm\sqrt{15}$$
$$x = 3 \pm \sqrt{15}$$

9. Writing as a complex number: $\sqrt{-9} = 3i$

10. Writing as a complex number: $\sqrt{-121} = 11i$

11. Writing as a complex number: $\sqrt{-72} = \sqrt{-36 \cdot 2} = 6i\sqrt{2}$

12. Writing as a complex number: $\sqrt{-18} = \sqrt{-9 \cdot 2} = 3i\sqrt{2}$

13. Combining the complex numbers: $(3i + 1) + (2 + 5i) = 3 + 8i$

14. Combining the complex numbers: $(6 - 2i) - (7 - 4i) = 6 - 2i - 7 + 4i = -1 + 2i$

15. Combining the complex numbers: $(2 + i)(2 - i) = 4 - i^2 = 4 - (-1) = 4 + 1 = 5$

16. Combining the complex numbers: $(3 + 2i)(1 + i) = 3 + 2i + 3i + 2i^2 = 3 + 5i - 2 = 1 + 5i$

17. Combining the complex numbers: $\dfrac{i}{3 - i} \cdot \dfrac{3 + i}{3 + i} = \dfrac{3i + i^2}{9 - i^2} = \dfrac{3i - 1}{9 + 1} = \dfrac{-1 + 3i}{10}$

18. Combining the complex numbers: $\dfrac{2 + i}{2 - i} \cdot \dfrac{2 + i}{2 + i} = \dfrac{4 + 2i + 2i + i^2}{4 - i^2} = \dfrac{4 + 4i - 1}{4 + 1} = \dfrac{3 + 4i}{5}$

19. Graphing the parabola:

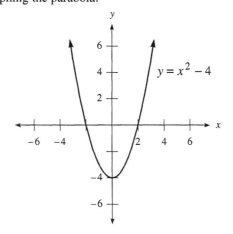

$y = x^2 - 4$

20. Graphing the parabola:

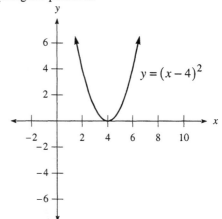

$y = (x - 4)^2$

21.   Graphing the parabola:

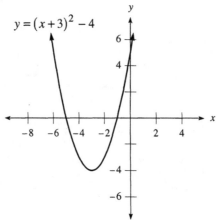

22.   First complete the square: $y = x^2 - 6x + 11 = \left(x^2 - 6x + 9\right) + 11 - 9 = (x-3)^2 + 2$

Graphing the parabola:

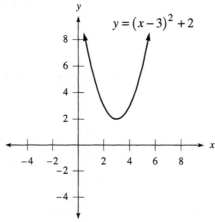